Praise for
THE LOST DINOSAURS OF EGYPT

"William Nothdurft tells this complicated, many-layered story in a plain style, with . . . earnestness. Fascinating material."
—*The New York Times Book Review*

"The sometimes perilous search for rare dinosaur bones can make for great adventures, as evidenced by *The Lost Dinosaurs of Egypt*. . . . [An] extraordinary paleontological discovery."
—*Discover*

"If you loved Indiana Jones, you'll adore this tale of two dinosaur hunters whose expeditions to Egypt, separated by nearly a century of warfare and mystery, brought to light what may have been the largest creature that ever walked the earth."
—Erik Larson, author of *The Devil in the White City*

"Fascinating. Nothdurft is a science writer who weaves a brilliant and colorful tale."
—*Newsday*

"This story is every bit as enthralling as the best adventure fiction. The A&E channel's production by Cosmos Studios . . . was excellent. See it when it is shown again, but, meanwhile, read the book."
—*The Roanoke Times*

"An engaging mix of history and desert drama; this Indiana Jones–type adventure is first-rate popular science."
—*Publishers Weekly*

"Told deftly and dramatically . . . The tale will keep you riveted as the Smith team unearths Stromer's lost dinosaurs and the 'giant near the sea' rises again."

"The Indiana Jones–style adventure of the American paleontology team's seven-week expedition through sandstorms and blazing heat [was] featured in *The Lost Dinosaurs of Egypt* [on A&E Network Television]. Josh Smith's gigantic dinosaur discovery on an Egyptian desert is the stuff dreams are made of."   —*The Recorder News*

"[*The Lost Dinosaurs of Egypt*] is interlaced with explanations of the principles of geology, geologic time, and paleontology. . . . A well-organized story of past and present that hones in on our fascination with dino hunters."   —*Booklist*

WILLIAM E. NOTHDURFT

Will Nothdurft is the author, co-author, or ghostwriter of nearly a dozen books of nonfiction, including the international award-winning expedition chronicle *Ghosts of Everest: The Search for Mallory and Irvine,* which was published in several languages. He lives in Seattle, Washington.

JOSHUA B. SMITH

Josh Smith holds a B.Sc. in environmental geology from the University of Massachusetts at Amherst (1994) and an M.Sc. in geology from the University of Pennsylvania (1997). He is scheduled to receive a Ph.D. in paleontology from the University of Pennsylvania in 2002. Josh has studied geology and paleontology in Massachusetts, Connecticut, Utah, Arizona, Colorado, Wyoming, Montana, Mexico, Puerto Rico, the Canadian High Arctic, Alberta, China, Argentina, and Egypt. He is the author or co-author of thirteen scientific publications and has directed the Bahariya Dinosaur Project since its inception in 1999. Josh currently lives in Cambridge, Massachusetts, with his wife, Jen.

MATTHEW C. LAMANNA

Matt Lamanna received a B.Sc. from Hobart College in 1997 and an M.Sc. from the University of Pennsylvania in 1999, and expects to receive his Ph.D.

from the University of Pennsylvania in 2002. He has discovered and excavated dinosaurs in Wyoming, Montana, Argentina, and Egypt. He lives in Philadelphia.

KEN LACOVARA

Ken Lacovara holds a Ph.D. in geology from the University of Delaware (1998), a master's degree from the University of Maryland, and a bachelor's degree from Rowan University. He has studied modern and ancient coastal systems around the world and is particularly interested in the coastal habitats of dinosaurs. A former professional drummer, Ken can still be seen on occasion playing in Philadelphia jazz clubs. He lives outside Philadelphia with his wife, Jean, and is an associate professor at Drexel University.

JASON C. POOLE

Jason Poole (aka "Chewie") is a fossil preparator, bone illustrator, field technician, and teacher at the Academy of Natural Sciences, where he focuses primarily on the Mesozoic world. Jason manages a staff of many volunteer and part-time preparators who support the work of the Bahariya Dinosaur Project and other projects in Montana, South America, and Pennsylvania. Jason resides in Philadelphia's East Oak Lane, where he grew up, and works in a museum that was an awe-inspiring influence in his young life. He spends his free time with a sketchbook and his family.

JENNIFER R. SMITH

Jen Smith received a B.A. from Harvard College (1996), and an M.Sc. (1998) and Ph.D. (2001) from the University of Pennsylvania. Her research involves the evolving relationship between people and their environment in the archaeological record. She firmly believes the geologist's adage, "Whoever sees the most rocks, wins," and is trying very hard to win. She currently resides in Cambridge, Massachusetts, with her husband, Josh.

THE LOST DINOSAURS OF EGYPT

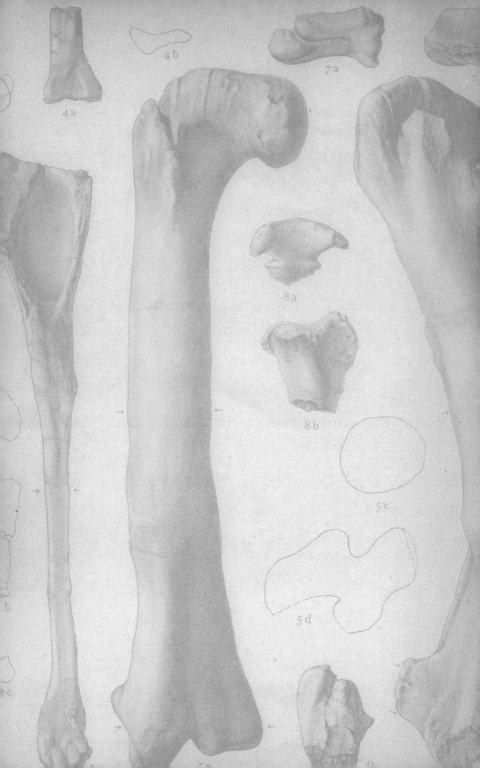

THE LOST
DINOSAURS
OF EGYPT

*The Astonishing and Unlikely True Story of One of
the Twentieth Century's Greatest Paleontological Discoveries*

WILLIAM NOTHDURFT

WITH JOSH SMITH, MATT LAMANNA,
KEN LACOVARA, JASON POOLE,
AND JEN SMITH

RANDOM HOUSE TRADE PAPERBACKS

NEW YORK

COSMOS
STUDIOS

Published in conjunction with Cosmos Studios

2003 Random House Trade Paperback Edition

Library of Congress Cataloging-in-Publication Data
Nothdurft, William.
The lost dinosaurs of Egypt / William Nothdurft with Josh Smith . . . [et al.].
p. cm.
ISBN 0-375-75979-4
1. Paralititan stromeri—Egypt—Bahariya Oasis. 2. Paleontology—Cretaceous. 3.
Dinosaurs—Egypt—Bahariya Oasis. 4. Paleontological excavations—Egypt—
Bahariya Oasis. 5. Stromer, Ernst. I. Smith, Josh. II. Title.

QE862.S3 N68 2002   567.9'0962—dc21   2002075172

Book design by Casey Hampton

CONTENTS

MAP OF EGYPT   ix

PROLOGUE: Death and Resurrection   3

ONE: Reaping the Whirlwind   7

TWO: The Bone-Hunting Aristocrat   21

THREE: Unearthing a Legend   46

FOUR: Dragomen, Fossils, and Fleas   66

FIVE: The Road to Bahariya   83

SIX: Finds and Losses   102

SEVEN: Sand, Wind, and Time   118

EIGHT: The Hill Near Death   138

NINE: Solving Stromer's Riddle   156

TEN: Lost World of the Lost Dinosaurs   180

EPILOGUE: Memorials   201

NOTES   207

BIBLIOGRAPHY   217

ACKNOWLEDGMENTS   223

INDEX   227

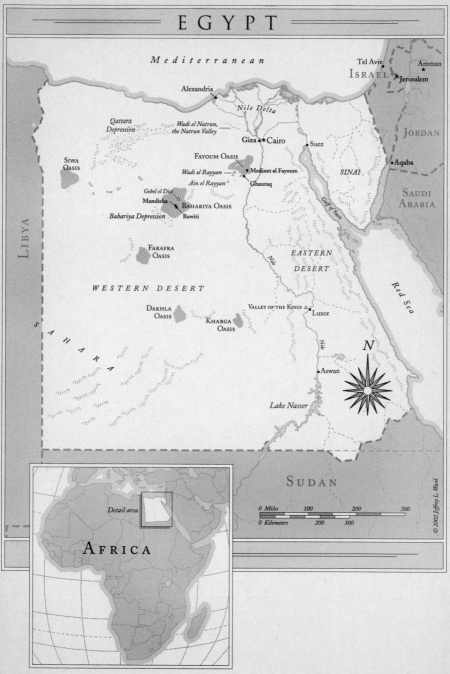

THE LOST DINOSAURS OF EGYPT

DEATH AND RESURRECTION

No one knows what brought the huge animal down. The life of a dinosaur had no shortage of perils. Even a dinosaur this big—more than 80 feet long and weighing perhaps 65 to 70 tons—was not immune. Something killed it: disease, injury, attack, possibly just old age. At some point its knees buckled and it dropped to the ground with a seismic thud, or perhaps a massive splash. It was almost certainly still alive at that moment—critically injured or racked with illness, perhaps, but still alive. Too weak to lift its head above the incoming tide, it may have drowned. It may have expired quietly, alone. It may have been surrounded by its fellow creatures, in much the same way elephants will gather around a stricken member of their herd. It may have been surrounded by far less sympathetic company. A predator, perhaps its attacker. Maybe more than one. The scent of death travels far. Ever opportunists, the predators may have begun dismembering the great animal while it was still struggling, still clinging to life.

Eventually the beast lost its battle against death. Almost immediately thereafter, possibly even before, when they knew it was safe to approach, the scavengers arrived—by land and air, even water—and began tearing apart the carcass. Soft tissue left behind by the larger scavengers was consumed by smaller ones, insects and bacteria, gradually decomposing into the soil or water beneath. The dead sustain the living: conservation of energy and matter.

In time, all that was left were the beast's great bones, scattered now, no longer part of an intact skeleton. Mud and sand drifting in and out on the tide collected around its remains. The process was so gentle that it did not disturb the bones much and merely buried them.

Bones are remarkably good at providing evidence of trouble, in both dinosaurs and humans. Bones preserve the signs of fractures and breaks even after they heal because the new material that fills in the cracks, called callus, is structurally different from the original. And even after new bone cells replace the callus, traces of the injury often remain. Disease can deform bones as well, twisting them, causing abnormal growths or altering their density or porosity. And the scars of tooth marks remain on bone even after predator and prey have been dead for tens of millions of years.

But over those millions of years, the bones can change. Very gradually, minerals dissolved in water percolating through the soil—silicate, calcium carbonate, iron oxide, calcium sulfate, and others—can permeate the porous structure of bone, filling in around and in some cases actually replacing the organic material of which the bone is composed. In the case of this particular dinosaur, the result of thousands of years of this exquisitely slow process was a certain kind of immortality for the fallen giant. It had become a fossil.

And that would normally be the end of the story. But in this case, it was just the beginning. On May 31, 2001, at a press conference in Philadelphia, the great beast was resurrected.

That same day, the June 1 issue of the prestigious magazine *Science* was released. In it was an article formally describing "A Giant Sauropod Dinosaur from an Upper Cretaceous Mangrove Deposit in Egypt." The article announced, in the typically arid prose of such journals:

> We describe a giant titanosaurid sauropod dinosaur discovered in coastal deposits in the Upper Cretaceous Bahariya Formation of

Egypt, a unit that has produced three *Tyrannosaurus*-sized theropods and numerous other vertebrate taxa. *Paralititan stromeri* is the first tetrapod reported from Bahariya since 1935. Its 1.69-meter-long [about 66.5 inches] humerus is longer than any known Cretaceous sauropod. The autochthonous scavenged skeleton was preserved in mangrove deposits, raising the possibility that titanosaurids and their predators habitually entered such environments.[1]

At the press conference, held on the campus of the University of Pennsylvania, Ann Druyan, founder and CEO of Cosmos Studios, stepped to the microphone before a large group of reporters and television cameras. To her left were five young scientists. She introduced the young man closest to her and he came to the microphone.

Josh Smith, a Penn doctoral candidate in paleontology, expressed his surprise at the number of reporters present and thanked them for coming. Then he told them a story. It was the story of a long forgotten German explorer and scientist who, almost a century earlier, had made not one but several astonishing dinosaur discoveries in, of all places, the Sahara Desert. It was a story about how the scientist lost them, and much more besides, through a series of crushingly tragic events.

It was also the story of a group of young contemporary scientists who believed they could resurrect the German's legacy and, at the same time, make a significant contribution to the world's understanding of what the planet looked like nearly a hundred million years ago.

In the Western Desert of Egypt in the winter of 2000, the group had succeeded, beyond their wildest imaginings, at both.

Smith introduced the other members of the Bahariya Dinosaur Project team: fellow paleontology doctoral student Matthew Lamanna and geology doctoral student Jennifer Smith, both from the University of Pennsylvania; Drexel University professor of engineering geology Kenneth Lacovara; and Jason Poole, head of the fossil preparation laboratory at Philadelphia's Academy of Natural Sciences. Smith also introduced two Penn faculty advisers who participated in the expedition: Dr. Robert Giegengack, experienced Egypt hand and chair of the Department of Earth and Environmental Science; and from the School of Veterinary Medicine, Dr. Peter Dodson, one of America's best-regarded vertebrate paleontologists.

Smith thanked three members of the team who were not present but whose participation had been critical to the success of the expedition: their Egyptian collaborators Yousry Attia, geologist and curator of the Egyptian Geological Museum, and Medhat Said Abdelghani and Yassir Abdelrazik, members of the museum's paleontology staff.

Then Smith introduced a bone.

Between the group and the audience was a long, heavily reinforced table with a large object covered by a plain white sheet. Smith and Jason Poole stepped up to the table, removed the sheet, and revealed a reddish-brown dinosaur bone of stunning immensity. A murmur went around the room; camera lenses zoomed. Then came the questions. The press conference went into overtime. But one question in particular captured the moment:

"So give us an idea of how big this animal was, say, compared to an elephant."

Matt Lamanna stepped to the microphone, laughing, and said: "This animal was as big as an entire *herd* of elephants."

The next day this lost dinosaur from Egypt was front-page news around the world. Josh Smith and Matt Lamanna would spend the next three weeks on the phone and in radio and television studios describing the great beast and the expedition that discovered it.

But as is so often the case, the media missed entirely the fact that the dinosaur, as big as it was, was only a small part of the story of the Lost Dinosaurs of Egypt.

This is the whole story.

ONE ══

REAPING THE WHIRLWIND

The second extinction of the dinosaurs from the Bahariya Oasis began shortly after midnight. It came from the sky. It began with a barely discernible disturbance in the air, a distant rumble that insinuated itself into the quiet of the night and quickly grew in intensity to a deafening roar. Then, suddenly, the sound became sight and the dark became light as the sky itself became fire. Moments later the roaring was punctuated by a stunning explosion that shattered the still night air. Then another. Then dozens more, until the earth shook and the ground split. Almost immediately, the sound and light became smell—the smell of burning, the singed stink of death. Screams rent the night, and soon the living became the dead.

There have been roughly a dozen mass extinctions during the history of life on Earth, five of them so severe and all-encompassing that they killed off vast numbers of living things. One was so catastrophic that it came close to ending life altogether. Indeed, all of the species alive today represent only 1 percent of all the life that has ever lived during the Earth's history. The other 99 percent have long since perished.[1] By far the worst of the mass extinctions occurred an estimated 245 million years ago and took several million years to run its course. But though it was gradual, it was also exceptionally deadly. Scientists believe fully 95 percent of all the forms of plant and animal life in the seas at that time were likely eliminated. Though the cause is still hotly debated, many scientists believe that the consolidation of all of the continents then in existence into a single landmass—called Pangaea—caused sea levels to fall, the land to heat, and the ocean to stagnate. In this scenario, carbon dioxide levels rose, the heat increased, oxygen levels in the ocean plummeted. Slowly but surely, life in nearly all its forms suffocated to death.[2] All we know about the creatures that vanished is what they left behind, their fossilized remains—petrified plantlike stems and calices of sea-dwelling crinoids, limy corals, bits of ammonite shell, skeletons of certain kinds of fish, tiny seagoing creatures.

But extinctions can also occur with cataclysmic suddenness. The age of the dinosaurs, those massive reptiles that ruled the Earth for more than 165 million years, appears to have ended abruptly, in geological terms, roughly 65 million years ago. To this day, no one knows why. One theory, intriguing though not widely accepted, points to the fact that this was a period of intense volcanic activity in many places on the Earth's crust. Perhaps the most spectacular eruption occurred in what is now southern India. There, between 66 and 68 million years ago, the Earth cleaved apart, spewing what may have been as much as 48,000 cubic *miles* of lava over an area of more than 772,000 square miles,[3] an area roughly three quarters the size of the entire American West. The remnant of this event is a formation known to geologists as the Deccan Traps.[4] The consequences of an eruption of this scale could have been appalling: Immense quantities of dust and ash would have been flung into the upper atmosphere and, in a matter of weeks, would have darkened the sky everywhere on the globe. In time, starved of light, plants would have shriveled and died. Animals that lived on plants would have followed, and animals that lived on other animals would, in turn, have followed them. What may have happened next is uncertain. The sulfurous air could have reduced temperatures sharply worldwide. Alternatively, the death of plants on land, and algae in the seas, may have caused carbon dioxide levels in the air to skyrocket, creating a massive greenhouse effect.[5] In no time at all, geologically speaking—perhaps only a few thousand years—the diversity of life on Earth would have been drastically reduced.

That is one theory. Scientists from the University of California at Berkeley and Lawrence Livermore National Laboratory proposed another, more frightening one: that the great age of dinosaurs was terminated by the impact of an object plummeting from space. Examining rocks in Gubbio, Italy, the scientists found surprisingly high levels of a rare element called iridium in a narrow band of rock that dated back 65 million years. Iridium does not occur normally in the Earth's crust in such concentrations: Most arrives from space through the gentle rain of cosmic dust and the somewhat less gentle arrival of small meteorites and asteroids. The accumulation of this element has been fairly consistent throughout time. But in this particular layer, the element appeared in the rock at a concentration equal to all the iridium that had been deposited in the preceding half million years![6] In 1980 researchers felt confident enough to make an announcement that was quite literally earth-shattering: 65 million years ago, they explained, an asteroid or comet roughly the size of Mount Everest struck the Earth at a speed of more than 22,000 miles per hour, creating an explosion 10,000 times more powerful than if all the nuclear bombs that exist today had gone off at once. The impact vaporized the comet or asteroid and spread iridium—and destruction—across a great swath of the Earth's surface, in roughly the same manner and to the same effect as the Deccan Trap eruptions.[7] The impact theory was strongly supported a decade later by the discovery of a crater, one of the largest yet discovered. Between 100 and 125 miles in diameter, it was found beneath the Yucatán Peninsula and the Caribbean Sea. Its date of origin? Roughly 65 million years ago. Other craters of similar age also have been discovered.[8]

So, which phenomenon caused the disappearance of virtually every single dinosaur on Earth? Maybe both, and other events as well.[9] It may well be that the age of dinosaurs was, in both ecological and evolutionary terms, an immense house of cards—intact but extraordinarily fragile. Or it may be more like the straw that broke the camel's back; as one scientist puts it, "Things got bad, then they got worse."[10]

Although similar in effect, the second extinction of the dinosaurs of the Bahariya Oasis, which occurred less than a century ago, had a different cause altogether. This particular extinction was a product of neither terrestrial nor extraterrestrial geologic forces. This extinction was man-made.

Wing Commander G. Leonard Cheshire arrived at the Royal Air Force's aerodrome at Woodhall Spa on the morning of April 24, 1944, as the soft spring sunlight began burnishing the hazy, expansive landscape of eastern England. An American expatriate to England, the poet T. S. Eliot, once wrote that "April is the cruellest month," but in Lincolnshire it can be positively radiant, the grass impossibly green, fields of nearly black soil freshly plowed and planted, lanes replete with naturalized daffodils and hedgerows frothy with hawthorn blossoms. As flat as a snooker table and richly fertile, this area just south of the Lincolnshire Wolds, along with the adjoining reaches of Cambridgeshire, contains to this day some of Britain's finest farmland, producing a wide array of market vegetables and flowers for the country's industrial cities. But after the outbreak of World War II, the region's principal crop was aerodromes. Close to the coast, and therefore to Nazi Germany, the farm fields became airfields. RAF Woodhall Spa, with three runways forming a rough triangle, a pair of corrugated-iron hangars, and a scattering of thrown-together brick huts, was simply one of dozens of airfields scattered across the eastern counties. The pilots and officers were billeted in a hotel in town that had been requisitioned by the Air Ministry. They got to and from the airfield mostly by bicycle.

As he approached the flight briefing room, Leonard Cheshire was effectively a walking miracle. An RAF bomber pilot for nearly four years, Cheshire by now should have been dead. The RAF's losses through the first three years of the war had been staggering. On average, of every hundred crew members in Bomber Command, only twenty-seven survived. Losses for each sortie or bombing mission ranged from 5 to as high as 10 percent. A single tour of duty for a bomber pilot involved thirty sorties. Mathematically, at least, a pilot couldn't be expected to live through one complete tour of duty. Cheshire was well into his fourth. He was twenty-seven years old.

Cheshire was an unlikely ace. With his movie-star looks and a college career at Oxford that he freely admitted was distinguished more by carousing than achievement, he hardly seemed a candidate for greatness. One biography describes his college years as "a time of fast cars, reckless exploits, fantastic extravagance, mounting debts and shady associations."[11] A student of the law, he graduated with a second-class degree, but that would turn out to be of far less importance to Cheshire, and to England, than another skill he learned at school: the science and art of flying. Cheshire joined the university's Air Squadron in 1936, and the undeniable panache attached to flying suited him perfectly. He was commissioned in the RAF Volunteer Reserve in 1937, as war with Germany began to seem inevitable; he joined No. 102 Squadron in June 1940, immediately after completing his degree. And there Cheshire seemed to find himself at last, quickly demonstrating remarkable flying skills and strong but compassionate leadership ability. Combining what his fellow pilots described as an ice-cold brain and hair-raising flight tactics, Cheshire soon won the admiration of his crews and the respect of the leadership of Bomber Command. During the next four years he and his crews were assigned ever more difficult missions. Unlike most of his fellow pilots and squadron leaders, however, Cheshire always made it home—though sometimes only barely.

On this particular morning, April 24, 1944, Cheshire knew that this, his hundredth mission, had an importance that far exceeded any other in his flying career to date. Cheshire had become the critical weapon in a high-risk battle between not England and Germany but two senior officers of RAF Bomber Command. That the outcome of this night's sortie might substantially affect the success of the upcoming top-secret Allied invasion of France seemed, at that moment at least, secondary to the war that had been waged for months between No. 8 Bomb Group commander Air Vice Marshal Donald Bennett and No. 5 Bomb Group commander Air Vice Marshal Ralph Cochrane.

The first three years of the war had been difficult and sometimes disastrous for the RAF. Initially, its bomber force wasn't large enough to pose a significant threat to the Germans (even as late as mid-1941, the RAF had only seven hundred serviceable planes available on any given day).[12] In addition, the bombers they did have lacked the speed, range, power, or altitude capabilities needed to drop large numbers of bombs on targets within Germany. At the same time, the RAF didn't have the long-range fighters needed to escort and protect bombers from German fighters during daytime raids. As a result, Britain could conduct only nighttime raids—a perfectly reasonable strategy if the RAF had developed the navigational technology to guide its bombers to their targets effectively at night. But they had not. Operating in the early years essentially by dead reckoning—quite literally "in the dark" about their own location—the bombers more often than not were unable to find their targets, and they often failed even to hit the cities in which their targets lay. And the RAF's losses were brutal. During 1941 its aircrew losses were actually higher than the civilian losses at its German targets.[13]

The British command concluded that the only way to wage a successful air war was to build a massive number of heavy bombers, work furiously to improve navigation technologies, and then mount a sustained campaign of area-wide bombing raids to destroy not just German military matériel, but also the homes, morale, and lives of German civilians as well. On February 23, 1942, the Air Staff named Arthur Harris—soon to be nicknamed "Bomber Harris"—air chief marshal to carry out the policy. He was "a commander of coarse single-mindedness [who] had neither intellectual doubt nor moral scruple about the rightness of the area bombing policy and was to seek by every means—increasing bomber numbers, refining technical bombing aids, elaborating deception measures—to maximize its effectiveness."[14] As new planes began to enter service, Harris was able to put as many as a thousand bombers in the sky on a single night, creating firestorms in some German cities that reached in excess of 1,000 degrees Fahrenheit; even the asphalt pavements caught fire. It was Harris who is famously quoted as saying of the Germans, "They have sown the wind, and now they shall reap the whirlwind."

Although the RAF's technicians steadily developed better navigational technologies (only to have their advantage quickly countered by Germany's own technicians), bombing accuracy was still unacceptably poor. Internal RAF studies found that despite the tonnages of bombs dropped, targeting was still "wide or wild," and the bombs were having little effect on Germany's ability to wage war.

In the end it was human innovation that began to improve the effectiveness of the RAF's bombing campaign. That innovation was the birth of the Pathfinder Force. The Pathfinders, created within Air Vice Marshal Bennett's 8 Group on August 24, 1942, were a group of elite pilots and crews who were assigned to fly ahead of a bomber formation and mark the targets with flares. Flying Lancaster heavy bombers and, on occasion, light, high-speed Mosquito fighter-bombers, the Pathfinders traced the route across Europe, then dropped brilliant markers to guide the bomber streams that followed. Bomber Command chief Harris initially had opposed the creation of the Pathfinders, fearing it would siphon off his best crews and strip the RAF of its leaders, but their success was undeniable and eventually he relented. Each target marker contained sixty pyrotechnic flares equipped with barometric fuses designed to set off at predetermined altitudes. As the air war progressed, the flares were color-coded daily to prevent German decoy flares from drawing the incoming bombers off target. Though much improved, accuracy was still inadequate and, as a weapon of war, "Bomber Command remained more of a cudgel than a rapier."[15]

The plain fact was that area bombing was not slowing appreciably the flow of matériel to German troops. Consequently, in June 1943, the Allied Command, meeting in Casablanca, changed the bombing rules on Harris and ordered that the area bombing campaign be replaced with a more narrowly focused one aimed at destroying smaller and much more strategic targets: refineries, rail yards, submarine bases, airplane engine factories, and transportation hubs, among others. Harris vehemently opposed this new offensive, code-named Pointblank, and for the most part simply refused to alter his long-standing commitment to massive area bombing.

This was more than just stubbornness on his part. For one thing, the

Germans had repeatedly demonstrated an astonishing ability to hide, relocate, or simply rebuild military manufacturing factories destroyed by Allied bombers. Harris argued that "strategic" bombing would have little practical effect. Only massive and sustained bombing of entire cities would break the will of the German people, cripple the country's industrial capacity, and starve its troops of the means of waging war at the fronts. But Harris and his subordinate Bennett knew something else as well, something they were perhaps somewhat less willing to admit: that even with the Pathfinders marking targets, their nocturnal bombers could not achieve the kind of pinpoint accuracy required by Pointblank. Bennett in particular was convinced that the Pathfinders could never fly low enough to mark such targets precisely without being destroyed by ground defense forces.

Enter Air Vice Marshal Cochrane, commander of 5 Group. While Bennett's 8 Group had the Pathfinders Force, Cochrane's 5 Group had the No. 617 Squadron—the so-called Dambusters—whose daring low-level bombing in 1943 had destroyed the hydroelectric dams in the Ruhr Valley, crippling German industrial capacity there and capturing the imagination of the British people. Cochrane was convinced the problem with the Pathfinders Force was that they released their flares at too high an altitude; he argued that low-altitude target marking could dramatically improve bombing accuracy without increasing casualties. At a meeting on January 18, 1944, he presented his case formally to Harris. Bennett, who attended, was opposed, and assumed the matter was closed. But Cochrane lobbied relentlessly for a chance to demonstrate his theory. What's more, he had an ace in the hole: Leonard Cheshire, now wing commander of No. 617 Squadron, had in early 1944 successfully marked the Gnome and Rhone engine plants at Limoges, France, from an altitude of only two hundred feet. The rest of Cheshire's squadron, following close on his tail, had demolished the factories.[16]

Throughout the early months of 1944, Cheshire and his squadron worked continually to fine-tune their low-level bombing tactics until, by early March, they could promise that upwards of 60 percent of their bombs would fall within a hundred yards of the target, an unheard-of level of accuracy. On April 14, Harris not only gave Cochrane the go-ahead but went so far as to transfer Bennett's No. 627 Mosquito fighter-bomber squadron as well as his No. 83 and 97 Lancaster bomber squadrons to Cochrane's command. Bennett was so furious he had to be restrained from resigning his command.[17]

Cochrane had more on his mind, however, than simply showing off his best fliers or advancing his career. Planning for Overlord, the Allied invasion of Normandy, was well advanced. D Day was only weeks away. If low-level target marking was feasible under extreme combat conditions, if it could significantly improve the accuracy of bombing raids, then nighttime bombing of heavily fortified German installations in France might be accomplished without causing massive losses among French civilians. The strategic importance of this demonstration mission was very high indeed.

Having made his Faustian bargain, Cochrane now learned what it was to cost him: Harris wanted Munich.

In the spring of 1944, Munich was the second most heavily defended city in Germany, after Berlin. If that was not sobering enough, it was also located so deep within the country that it lay at the very limit of a Mosquito's round-trip fuel range. Assuming they met with no fuel-consuming headwinds, dropped their markers on their first try, and were not required to undertake many evasive maneuvers, Wing Commander Cheshire's target markers would have just fifteen minutes of fuel to spare when—and if—they returned to England.

Bomber Command chose Munich for reasons as much symbolic as strategic. Certainly destroying the rail center, the raid's principal objective, had strategic value. But Munich was also the birthplace and ceremonial home of the National Socialist German Workers' Party—the "premier city of the Nazi Movement."[18] Berlin may have been the administrative heart of the Third Reich, but Munich was its soul. Destroy it and you would destroy Nazism's birthplace and drive a symbolic stake into Hitler's heart. The RAF had attempted to do just that in the early months of the war, with catastrophic effect—not to Germany, but to England. It was this early unsuccessful raid on Munich that resulted in the retaliatory raid by Germany which destroyed Coventry, England. Perhaps Harris had been waiting all this time to get even.

And so, sometime around midday on April 24, 1944, 617 Squadron Wing Commander Leonard Cheshire, the marker leader for the mission, taxied his Mosquito to the end of the runway at Woodhall Spa aerodrome, raced down the end of the airstrip, and lifted into the air. Three other Mosquitoes followed suit, piloted respectively by Squadron Leader David J. Shannon as deputy leader and Flight Lieutenants G. E. Fawke and R.S.D. Kearns as assistant deputies. Though Cheshire had spent most of his years as a pilot at the controls of Lancasters, the deHavilland-designed Mosquito was uniquely suited to this mission. Lightweight—manufactured, in fact, largely of plywood—and fitted with a pair of powerful and reliable Rolls-Royce Merlin engines, the Mosquito could fly faster and higher and maneuver better than any other plane in the war at that time.

Together, the four planes turned south and made for Kent, in southeastern England. There, at RAF Manston aerodrome, they would top up their fuel tanks and await the order to take off again, this time for Munich.

Even before darkness had fallen, from airfields scattered across the countryside around the cathedral city of Lincoln, bomb-laden Munich-bound Lancasters lumbered into the air. By this stage of the war the formation pattern was well established: The bombers rose and circled over eastern England, massing and eventually forming a dense bomber stream before turning east toward the European coast and their target. In all, the raid on Munich would involve 244 Lancaster heavy bombers and 16 Mosquitoes.[19] By 1944 the Lancaster, with its four Merlin engines and its immense bomb-carrying capacity (up to twenty-two thousand pounds of bombs by the end of the war), had become the RAF's "Shining Sword." Lancasters had the range, speed, altitude, and sheer flyability that earlier bombers had lacked. In addition, with hydraulic gunner turrets fore, aft, and amidships, Lancasters were formidably able to defend themselves.[20] They were tough, resilient flying machines; it was not uncommon for ground crews to hear, long after the main force had returned home, a Lancaster straining through the early-morning sky and see one of their crippled charges skim the hedgerows at the end of the airfield to touch down safely, if awkwardly, with only two engines functioning.

The main bomber force took off in daylight because Bomber Command had devised an elaborate ruse to throw off the German defense forces. Instead of turning east toward Germany, the Lancaster force flew south, crossing the English Channel near Southampton, passing over the Normandy coast at Deauville, continuing on this southerly course to Romorantin, just east of the grand châteaus of the Loire Valley. There the bombers turned to the southeast and flew high above the vineyards of Burgundy toward Geneva. To German trackers, their route was unmistakable: heading straight for Milan, Italy. And indeed a small force of sixteen Lancasters did exactly that, scattering flares and a cloud of tinfoil strips (called "window" by the RAF) over the city to trick radar installations there into believing a huge force of bombers had arrived. In fact, the main force had turned due east just beyond Geneva and then north, heading directly for Munich.

After German command recognized the false alarm in Milan, they had new trouble on their hands: Another force of 637 RAF bombers had mounted a simultaneous raid on the industrial city of Karlsruhe in southern Germany, near the French border,[21] and German defensive fighter planes raced back to defend that city. By the time the German command realized another force was headed for Munich it was too late; the fighters were effectively out of fuel.[22]

As these various feints and evasions got under way, Cheshire's four Mosquito target markers streaked through the sky on a ramrod-straight route from Kent, over Belgium and across southern Germany to Munich. The weather was terrible. The Mosquitoes lifted off from RAF Manston several hours after the main bomber force and immediately ran into dense cloud cover. They managed to stay on course and broke out of the cloud cover near Augsberg, just northwest of Munich, where they were immediately picked up by ground defenses, plastered with searchlight cones, and subjected to withering flak attacks. Miraculously, they reached Munich shortly after midnight, on time and unscathed. Here some three hundred searchlight beams groped for their planes in the night and bursts of lethal radar-guided flak were heavy, in part because an advance group of Pathfinder Lancasters from the main force had just released a stream of flares high above the city to enable Cheshire and his fliers to find their marks.

Cheshire was the first of the target markers to arrive. Looking down at the brilliantly illuminated city from an altitude of five thousand feet, he saw his aiming point directly beneath him—a Gestapo building just east of the raid's strategic target, Munich's central railway station and its switchyards. A proper approach would have meant crossing the city and returning, which, given his fuel limitations, he could not afford. Immediately and instinctively, he dropped one wingtip and put the Mosquito into a vertical dive he knew would exceed the safe speed limits of the plane. Only a few hundred feet from the ground he began to pull up, releasing his markers and placing them virtually on the rooftop of his target. Struggling to retain consciousness in the enormous G-force created as he reversed the dive, he put the plane into a steep climb through the flak clouds. He called in the other three Mosquitoes to mark their targets. Circling the city at just a thousand feet, Cheshire, acting as the operation's master bomber, summoned the main bomber force to the markers. They came in two waves, separated by a few minutes.

Cheshire's own situation was now extremely perilous. His plane was illuminated by light from above and below and subjected to intense ground attack. Above him, waves of Lancasters passed high above the city. One by one, they opened their bomb bays. The sky filled with bombs and incendiaries. Moments later, explosions traced lines of fiery destruction across the center of Munich.

With fuel already dangerously low and flak explosions thundering around them, Cheshire finally took the four Mosquitoes home. Searchlights and flak followed them, and Cheshire was forced to execute violent and fuel-costly evasive maneuvers. Twisting through the sky, they finally outran enemy fire. With luck, they would have just enough fuel to make it back to RAF Manston airfield. And luck was with them, at least until they crossed the English coast and began to descend. Ahead, the field was illuminated oddly, as if it was having electrical trouble. It was Cheshire who realized what was up: A lone German fighter plane was shooting up the airfield. After a few more evasive moves, luck returned: The fighter withdrew, and the four Mosquitoes landed safely. They had little more than fumes left in their fuel tanks.

Munich in 1944 was a city of broad squares and grand public buildings built in the sort of neoclassical monumental architectural style so characteristic of the era. One such building, the Alte Akademie, occupied nearly an entire city block. Its main entrance was on Neuhauserstrasse, roughly a half mile from the *bahnhof.*

The Alte Akademie was, in 1944, the home of the Bavarian State Collection of Paleontology and Historical Geology, one of Germany's and the world's most respected centers of scientific research. Within its high-ceilinged halls was housed, among many other antiquities, an extraordinary collection—the 95-million-year-old bones of four huge dinosaurs entirely new to the world of science that had been found some years earlier in the barren wastes of the Western Desert of Egypt. The fossils had been found by a Munich-based German explorer and pioneering paleontologist near an ancient oasis known to desert dwellers for centuries as El Bahria, the Bahariya Oasis.

The RAF bombing of Munich ended at one-forty A.M. on the morning of April 25. At dawn, the city's residents would discover that the central railway station—the kind of strategic target 8 Group Vice Marshal Bennett had refused to believe could be safely marked at low altitude—was now a mass of twisted steel and burning rubble.

Despite Cheshire's exceptional marking, however, there was collateral damage beyond the station. With more than two hundred Lancasters dropping hundreds of bombs, it was inevitable. More than seven thousand buildings in the vicinity of the station were in flames. One of them was the Alte Akademie. As the wan daylight filtered through the smoke, it was clear all that was left of the Akademie was a hollow masonry shell. Sometime after midnight an RAF bomb had slammed through the roof and set the museum ablaze. The collection was destroyed.

The dinosaurs never knew what hit them.

TWO

THE BONE-HUNTING
ARISTOCRAT

Anyone taking a stroll on the upper deck of the Lloyd's steamship *Cleopatra* on the morning of Wednesday, November 9, 1910, could not fail to note the gentleman standing at the railing, nor that he was thoroughly disgusted with the situation in which he found himself. A small bespectacled man, carefully groomed with a neatly trimmed Vandyke beard, he was dressed—as he had been throughout the voyage—with a certain studied precision. On this day he wore a white shirt, tie, dark trousers, handmade shoes, and a thigh-length wool coat buttoned, as was the fashion of the day, just once, at the breastbone. And because he was a gentleman, he of course wore a hat. He stood ramrod-straight, almost visibly vibrating with impatience, his delicate hands clasped tightly behind his back.

Before him sprawled the teeming port of Alexandria, Egypt. Even now, in November, and with the day still young, the heat was rising. All along the ramshackle docks, hundreds of laborers dressed in what seemed to be little more than rags scurried about, backs bent under heavy bundles, loading and unloading steam and sailing vessels of every description and condition. As the morning warmed, the air was redolent of tarred wharf timbers, horse and donkey dung, fetid seawater and sweat. Beyond the docks lay a warren of narrow alleys, tawny mud-brick warehouses and primitive dwellings, the flat-roofed scene punctuated here and there with taller municipal buildings of a decidedly French colonial design and, taller still, the minarets of Muslim mosques. The entire scene was bathed in that peculiar quality of light the man had come to associate with Egypt, a pale golden haze—part sunlight, part wood and coal smoke, part the ever present fine-grained dust of the Sahara: carried on the wind for hundreds of miles, filling the air and giving it texture before finally settling in shallow drifts in the nooks and crannies of the ancient port city. On any other day, this prospect, of which he had become curiously fond over the years, would have filled him with pleasure.

But not this day. Nor even the day before. The gentleman had had enough of this view. The ship had docked two days earlier, only to be slapped with a quarantine when the captain revealed that someone belowdecks in third class had a disease he suspected was cholera. Under police guard, the cargo had been unloaded—huge sacks of sugar from Hungary, bags of apples, crates of wine—but the passengers had remained on board. As the hours crawled by, a doctor of uncertain qualification moved slowly among the passengers, checking their pulses and making them stick out their tongues. Trained in medicine himself, the gentleman in first class was not impressed. He was not the sort of fellow who suffered fools easily, and now he seemed beset by them. He was sick of the delays, of the contradictory assurances and apologies of the captain, of the simultaneous officiousness and inefficiency of the port police, and of the endless, pointless waiting. He had important work to do and a limited time within which to accomplish it.

The impatient gentleman's name was Ernst Freiherr Stromer von Reichenbach. Any fellow German on the boat would immediately have taken note of the "Freiherr," a word that translates roughly as "baron" in English and signaled instantly the gentleman's aristocratic standing. In fact, members of his family had been pillars of his home city of Nuremberg, in what was then the kingdom of Bavaria, since the fifteenth century. His ancestors had been courtiers, lawyers, judges, architects, scientists, and political leaders. The gentleman's own father had been the city's mayor. Stromer himself was a scientist, an associate professor at the highly respected University of Munich and a man with a rising reputation in his field. Though he had originally studied medicine and science, his field now was geology and paleontology. His specialty was the paleontology of the Western Desert of Egypt. He was just past his fortieth birthday, and this was his third expedition to that remote and inhospitable part of Africa.

Notwithstanding the present unpleasantness, the voyage so far had been less than salubrious. The suitably, if somewhat unimaginatively, named *Cleopatra* had sailed from Trieste at noon on the previous Thursday. Stromer had arrived the night before, having traveled by train from Germany along a route that had taken him from Munich through Mühldorf and Salzburg, then south through spectacular mountain scenery bright with the colors of autumn, before finally arriving at the port on Wednesday evening. To save money, he had taken a third-class train ticket and, for the single evening he was to spend in Trieste, booked himself into an inexpensive hotel, only to find the bed verminous. Though it pained him to spend the money, that night's experience had caused him to book a first-class cabin (though one he shared with another passenger) for the voyage to Egypt.

For a few hours the *Cleopatra* steamed southeast along the Istrian coast of what is now Slovenia, then through the coastal islands of Croatia before heading out across the Adriatic toward the heel of Italy's boot. The sail through the islands had been splendid, the low slant of the November sun coating the islands with gold. But later that afternoon the sea turned rough, and Stromer spent the time on a deck chair trying, with typical German determination, to control his unruly stomach. Unsuccessful, he skipped dinner altogether, retired to his cabin, and spent the night seasick. To make matters worse, at one-thirty A.M. Stromer discovered that his first-class stateroom, which he had assumed would guarantee him a modicum of comfort and peace, was directly below the galley, from which there arose a clattering that would awaken him in the wee hours of each morning throughout the trip.

Still, the second day dawned warm and sunny, with a calm sea, and Stromer regained his stomach and his good humor as the ship sailed south under a crisp blue sky. The ship docked briefly at Brindisi at three P.M., then turned away from Italy and made for the southwestern coast of Greece. Having once again slept miserably because of the noise, Stromer rose early on Saturday morning to see first the sunbaked Ionian island of Kefallinia and, later, Zakinthos, passing just off the starboard bow as the boat steamed south along the inside passage toward the rugged Peloponnesian coast. By sunset the ship had reached the Gulf of Messini, and under the light of the moon, it slipped past Crete and headed out into the Mediterranean. The next day, as the steamer crossed the sea toward Africa, Stromer lounged on the deck in the sun, passing time with the other first-class passengers. That night, in what would turn out to be the vain hope of a restful sleep, he turned in early. The next morning, Monday, November 7, he was up at dawn to scan the horizon for the approaching North African coast and the port of Alexandria. It was as the *Cleopatra* was easing into its berth on Tuesday afternoon that the captain revealed news of the quarantine.

In some respects, the dockside delay was the least of Stromer's problems. On this expedition, he was to be accompanied by another scientist whom he identifies in his meticulously kept field journals only as "Dr. Leuchs."[1] However, the relationship between them began badly and continued to deteriorate. For one thing, Stromer (at this time still a bachelor) was astonished to find, when he boarded the *Cleopatra,* that Leuchs had arrived on the expedition with his wife. Furious at the extent to which this would hamper what would certainly be a physically difficult journey—therefore compromising his scientific objectives— Stromer nonetheless swallowed his irritation, remained cordial, and even attempted to teach Leuchs the rudiments of Arabic only to find him a thoroughly uninterested student. To make matters worse, as Stromer details in his journal in the spidery Sütterlin handwriting that was even then falling from favor in Germany, both Dr. and Frau Leuchs treated him rudely when they considered him at all. It was, in sum, an intolerable situation. For Stromer, being kept prisoner aboard the *Cleopatra* was made doubly trying because it forced him to remain in close proximity with the doctor and his distant, unfriendly wife.[2] Indeed, it would be hard to know which prospect he found more troubling that morning: spending more time on board the ship or spending more time in Egypt with "*Dr. Leuchs und Frau.*" Neither alternative was even remotely attractive.

At last, on the ninth, the captain announced that the passengers would be released. After a night ashore at a hotel of dubious quality arranged through a local hired by Leuchs, Stromer and his companions finally boarded a train at noon the next day, rattled along the western edge of the lush Nile Delta, and arrived in Cairo at three P.M.

The Cairo they entered as they disembarked from the train that afternoon was a city that seemed in the grip of chaos. Not that there was any crisis afoot; chaos was simply endemic in Cairo in 1910. It had been thus for twenty-five centuries, as Egypt had been under some form of constantly shifting foreign control since 525 B.C.[3]

Europe's influence in Egypt began in 1798, when Napoléon Bonaparte, with an eye to cutting off the British from their growing holdings in India and the Far East, landed an invasion force at Alexandria, advanced up the Nile to Cairo, and crushed the local chieftains, the Mameluke beys, at the Battle of the Pyramids. He declared himself an admirer of the Muslim faith and a disciple of Muhammad, and claimed his goal was to return Egypt to the Ottoman sultan's control. Egypt was nominally a part of the Ottoman Empire, but the sultan had never been able to win the full support of the Mamelukes. But neither side trusted Napoléon,[4] and France abandoned its claim to Egypt in 1801, to be replaced immediately by a joint Anglo-Ottoman expeditionary force.

Napoléon's invasion and the subsequent French occupation did have three immediate and lasting effects. First, it awakened his enemies, the British, to the strategic importance of Egypt and the Ottoman Middle East to their colonial aspirations. Second, somewhat ironically, it turned many leading Egyptians into avid Francophiles. Forward-looking Egyptians and representatives of the Ottoman sultanate in Egypt adopted French customs, sent their children to school in France, and built homes and official buildings in the French style of the era. (Many of these elegant buildings remained when Stromer arrived in Cairo on his first expedition in 1901.[5]) Third, a corps of French intellectuals was dispatched to study Egyptian culture and history. They brought back detailed accounts of the monuments and wonders of Egypt, sparking an obsession in Europe with all things Egyptian.

But there was another, even more lasting effect. An Albanian Muslim, nominally in the service of the sultan but a man of dubious loyalty to anyone but himself, arrived in Egypt with the Anglo-Ottoman force. In 1805 he stepped into the void left by Napoléon and the weakened Mameluke beys and seized power. His name was Muhammad Ali. He and his heirs would rule Egypt for a century and a half.

Muhammad Ali was succeeded in 1849 by his grandson Abbas,[6] who was an admirer of the British and granted to Britain the concession to build a railroad—the first in Africa or Asia—from Alexandria to Cairo. Completed in 1851, the railroad dramatically strengthened Britain's foothold in the region and simultaneously eased its imperial passage to India. Abbas's successor, his uncle Said, was a French loyalist and had long been friends with the French engineer Ferdinand-Marie de Lesseps, to whom, in 1856, he awarded the right to build a canal from the Mediterranean to the Gulf of Suez—the Suez Canal.

It was Said and his successor, Ismail, one of Muhammad Ali's younger sons, who moved Egypt into the modern era and, ultimately, bankruptcy. The two put Egypt into a rapid campaign of modernization, building new railways, roads, telegraph lines, canals, dams, bridges, and municipal infrastructure at a pace that prompted a correspondent of *The Times* of London to write, "Egypt is a marvelous instance of progress. She has advanced as much in seventy years as many other countries have done in five hundred."[7] Much of this progress, however, was funded by European loans carrying ruinously high interest rates. And because the Ottoman sultanate, which still ruled Egypt, had leased out much of the country's economic activity to European commercial concessionaires at very low rates, the Egyptian leaders were strapped for funds with which to cover the state's debts.

In a desperate attempt to reduce the debt load and stave off bankruptcy, in 1875 Ismail sold Britain his 44 percent interest in the Suez Canal for a mere £4 million, making Britain the principal stockholder (thus infuriating the French). But even this could not stem the cresting tide of debt; Egypt formally declared bankruptcy a year later.[8] Thus, what Britain and France had failed to do militarily or politically, they achieved instead financially: Egypt was ruled now by its creditors, through a joint Anglo-French debt commission.

When, a few years later, a nationalist uprising threatened to upset the European control of Egypt through its puppet—Ismail's successor, Tewfik—most of the European powers were surprisingly unwilling to undertake a military action against the uprising. Britain, with perhaps the most at stake, acted on its own in crushing the rebel forces. From that point on, Britain ruled Egypt as a protectorate. Six years before Stromer and his companions arrived in Cairo, the 1904 Entente Cordiale between France and Britain solemnized the arrangement.[9] In the agreement, France received Morocco as its protectorate in return for ceding Egypt to Britain. It was to these British authorities that Stromer would have to appeal for the permits he needed to enter the Western Desert.

Despite its crowds, hawkers, hustlers, noise, and furious activity, Cairo in 1910 was still only Egypt's second city, after Alexandria. But change was in the air. Some roads had been macadamized. Formal government buildings were well established, and construction activity was everywhere. Though horses, donkeys, and camels filled the streets and pedestrians picked their way around them and their droppings, there were also some five hundred automobiles that tried, with limited success, to squeeze their way through the narrow streets and milling throngs—a harbinger of the vehicular anarchy that rules the city's streets today.[10]

British domination of Egypt notwithstanding, Stromer's social calendar in Cairo during November 1910 demonstrates a substantial German presence there in the years before World War I. The Germans played a significant role in exploring and mapping the Western Desert. Georg Schweinfurth, for example, founded the Royal Geographical Society of Egypt in the late 1800s. Schweinfurth, also president of the Institute of Egypt, was an archaeologist, geologist, and botanist who worked in Egypt, and particularly in the Fayoum Oasis southwest of Cairo, for more than thirty years.[11] Schweinfurth's discoveries of mammal and other fossils in the Fayoum had led Stromer to Egypt on his first expedition.

Although other Europeans had explored portions of North Africa, it was the German geographer Gerhard Rohlfs who mounted the most comprehensive expeditions, crisscrossing the northern portion of the continent six times in the latter half of the nineteenth century and amassing an immense amount of information on the region. In 1873 Rohlfs secured Ismail's sponsorship for an expedition to explore and document the "unknown territories of Egypt" in the Western Desert. While Rohlfs's objectives were scientific, Ismail's were fundamentally economic: He asked Rohlfs to search for the rumored old riverbed of the Nile and to determine whether agriculture was possible in the Western Desert.[12]

A massive undertaking that required months to plan and equip, the Rohlfs expedition was, among other firsts, the first to produce a reliable though still incomplete geological map of the Western Desert. It was compiled and ultimately published by one of the scientists who accompanied Rohlfs, the famous German geologist Karl Alfred von Zittel. A few years later von Zittel, based at the University of Munich, would serve as mentor and thesis adviser to a promising young student of geology and paleontology. The student's name was Ernst Stromer.

After checking in to his Cairo hotel, Stromer went directly to the post office and found a letter of welcome waiting for him from the director of the Geological Survey of Egypt. While the older Royal Geological Society had been created by a German, the Geological Survey was founded in 1896 by the British, primarily to map Egypt's boundaries in order to defend them from increasingly troublesome attacks from Sudan, in the south. Both institutions had a history of cooperation by the time Stromer arrived, and he had established good working relationships with each during his earlier expeditions to the Fayoum Oasis and Wadi el Natrun, the Natrun Valley, during the winters of 1901–1902 and 1903–1904.

Stromer was, by all accounts, a man who observed the formalities, and the second thing he did that first afternoon in Cairo was pay a visit to the office of Georg Steindorff, a noted German Egyptologist. The visit would have been both a matter of courtesy to the senior scientist and part of planning the current expedition. In 1901 Steindorff had visited several of the oases of the Western Desert,[13] including the one that would occupy much of Stromer's time on this expedition and much of the rest of his life—Bahariya—and he needed Steindorff's counsel.

Meanwhile, in between meetings, and whenever Leuchs and his wife were not out sight-seeing somewhere on their own, Stromer introduced them to his German and British associates and acquaintances, including the great Schweinfurth, with whom he remained close friends. He also invited the couple to dine with him at the German tennis club and attempted to expose them to the range of artistic and cultural attractions in Cairo. To Stromer's growing annoyance, the couple showed neither interest nor gratitude.

It was with what one senses must have been relief that, on November 14, Stromer went off to meet with John Ball, founder of the Desert Survey Department of the Geological Survey of Egypt. Much of what is known about the Western Desert today is due in large part to the work of Ball and his early colleagues at the survey: H. G. Lyons, its first director, Hugh Beadnell, William Fraser Hume, and Thomas Barron. Once again, Stromer's visit was both formal and practical, for just that year the Survey had published the first topographic map of Egypt, in six large sheets, and was even then finishing up a geological map that would be published in 1911. Both resources would have been invaluable to Stromer for his upcoming expedition to Bahariya, one of the less well known areas of the Western Desert.

On the morning of November 15, after yet another unpleasant evening with the Leuchses, Stromer was worried about two matters. The first, and in many respects the most important, was a missing person, one Richard Markgraf. Markgraf, a somewhat mysterious figure about whom little is known, was one of those people of European descent, common even today, who fall in love with the deserts of North Africa and will do anything, often living hand-to-mouth, to stay there. Originally from Bohemia, Markgraf lived in the village of Sinnuris in the Fayoum Oasis, just south and slightly west of Cairo. He was a commercial collector of fossils and other natural curiosities who sold the items he found to paleontologists and museums, principally in Europe. Markgraf had worked with earlier fossil hunters in the Fayoum, among them the German paleontologist Eberhard Fraas (who would later make remarkable discoveries of Late Jurassic dinosaurs in East Africa) and Henry Fairfield Osborn, from the American Museum of Natural History.[14] There is no record of how Markgraf came to Stromer's attention, although several of Stromer's friends in Egypt, including Schweinfurth and the English geologist Hugh Beadnell, would both certainly have known Markgraf and could well have recommended him to Stromer. What is known is that they met during Stromer's first visit to the Fayoum in the winter of 1901–1902, and they got along well: Markgraf served as Stromer's *sammler,* or fossil collector, for a decade and a half, and became his friend. By all accounts Markgraf had a remarkable ability to recognize and carefully excavate significant fossils. At least three fossil animals are known to have been named in his honor, the early primates *Moeripithecus markgrafi*[15] and *Libypithecus markgrafi,* and a fish, *Markgrafia libica.*

But Markgraf often was not well. It is unclear from Stromer's journals whether the cause was malaria, still a chronic danger in some oases during the spring and early summer, or intestinal bleeding, perhaps from typhoid or chronic amebic dysentery (there are references to both kinds of symptoms). On this particular morning, Stromer was worried that Markgraf was having one of his spells. Stromer desperately needed Markgraf to help organize and equip the expedition, on which the collector had agreed to accompany Stromer. So far, however, he had not appeared and time was running out. Attempts to reach him had failed.

Stromer's second worry was chronic: how to address the fact that a joint expedition with Leuchs, and now his wife, was rapidly becoming an intolerable notion. As it happened, Leuchs caught Stromer in the hallway of their pension, on his way to his room after breakfast, and shortly resolved the problem by starting a row. He attacked the stunned Stromer for "creating a scene" when the doctor had appeared aboard the *Cleopatra* with his wife, accused him of rudeness and of hating his wife and women in general, and complained that his wife found Cairo unbearably dull. No doubt at the limit of his very considerable self-control, his voice low but tight with tension, Stromer reminded the good doctor that he had told him even in Munich that the budget would not permit the addition of Frau Leuchs, pointed out that he had gone out of his way to introduce the doctor and his wife to friends in Cairo, and suggested that the desert would provide far less in the way of entertainments than Cairo. It is entirely unclear what Leuchs had expected of the Egyptian desert or what he had intended to do on the expedition, but in any case Stromer had had enough. He offered to provide Leuchs with half of the expedition's water containers and supplies so Leuchs could pursue his own journey, and even volunteered Markgraf as companion and guide. Leuchs rejected the assistance of Markgraf but accepted the supplies, and that was the end of it. Stiffly, Leuchs turned on his heel and stalked down the hall. But for an occasional mention in his later journals, Herr Doktor Leuchs and his wife vanished from the scene.

Stromer's other worry that morning similarly vanished the moment he inserted his key into the lock of his door, turned the handle, and entered his hotel room. There, sitting in a chair, was Richard Markgraf.

Stromer's plan for this season's expedition was in three parts. First, he and Markgraf would head northwest from Cairo to Wadi el Natrun, which lay roughly at the halfway point and some distance to the west of the Cairo-Alexandria rail line. After exploring that region for a few weeks, they would return to Cairo, replenish their supplies, and then head south to Luxor to explore the eastern slopes of the Nile Valley, where Stromer believed he would find rocks and fossils similar to those in the Natrun Valley. Finally, they would return again to Cairo, restock, acquire additional permits, and then turn southwest for the main focus of the expedition, far out in the Western Desert: the Bahariya Oasis.

The days prior to their departure were filled with preparations. The crates of supplies Stromer shipped to Cairo from the docks at Alexandria arrived, and he conducted a complete inventory. Among his most important possessions were three square metal boxes, sealed with screws, in which he would carry the expedition's water. Awkward and heavy—each box held 50 liters (13.2 gallons) and weighed 25 pounds empty and 125 pounds filled—these boxes were critical to the success of their journey. They were the only drinking water Stromer's little expeditionary force would have for long stretches of desert that had neither oases nor wells.[16] The boxes originally had been designed and manufactured in Germany for the Rohlfs expedition to the Western Desert more than thirty-five years earlier but were still the standard. On November 16 Stromer and Markgraf shopped for tents and additional supplies and, more important, tracked down one of the native Egyptians with whom they had worked before, a camel driver named Oraan. Stromer and Markgraf arranged for Oraan to secure the camels they would need for the trek into the desert. Then, on November 17, Stromer secured the final permissions from Egyptian and British authorities, including a vital permit to refill the water boxes at a well that he knew from his previous expedition would provide safe water.

Gaining permission to enter the desert was no longer as easy as it once had been. Even in 1910 tension was growing between Germany and Britain, and the two countries were wary of each other's activities anywhere on the globe. While their scientific communities still worked together easily in Egypt, the political authorities were a somewhat different matter. There was, after all, a history. Kaiser Wilhelm, as later became horribly evident, was a rabid German nationalist—this despite the fact that his own mother was a daughter of Queen Victoria's. He was also unabashedly expansionist. Only two years before Stromer's first visit to the Egyptian desert, the kaiser had gone to Damascus, to the tomb of the great Saladin, and had sworn that the Ottoman Empire would henceforth have Germany's protection—a pledge the British took as a none too subtle threat.

Thus it would not be at all surprising if the British authorities thought twice about permitting a German like Stromer to mount a new expedition into the desert. But here his investments in courtesy and re- lationship building paid off. He was by now well respected among not just the trusted German scientists who had long worked in Egypt, like Schweinfurth and Steindorff, but also by the British-led Geological Survey. He got his permits.

Papers finally in hand, Stromer began his fieldwork. He and Mark- graf took the tram from Cairo to Giza, where they joined their trusted camel driver, Oraan, and loaded their four camels. Then, according to Stromer's journal, at nine-forty in the morning on November 18, he, Markgraf, and Oraan began hiking across the Giza plateau, their little train of camels following behind.

As they marched northwest toward Wadi el Natrun—a valley named for the native natron salts that, centuries earlier, had been instrumental in mummification—Stromer had an explicit objective. His principal interest was discovering the fossils of early mammals in North Africa, both ma- rine mammals like early whales and sea cows, and land mammals. It was a passion driven by his conviction that mammals, including humans, had originated not in Europe or the northern continents but in Africa. This was not a popular notion. In the late nineteenth and even early twentieth centuries, it was still widely believed that mankind had originated where it now was clearly at its most advanced state of development, specifically Europe. Another theory, promulgated by the American (though Cana- dian by birth) scientist William Diller Matthew, argued that North America was the source of most groups of mammals.[17] The idea that human beings could have evolved somewhere else, somewhere "more primitive," was simply not to be entertained. The fact that most of the information collected to date had come from Europe and North America and that Africa and South America were virtually unexplored, and that this might have skewed the conclusion somewhat, seems not to have occurred to many people at the time.

In 1901, when Stromer first visited Egypt and the Fayoum Oasis in search of mammal fossils to support his belief, he was not the first to look there. Only a few years earlier, his friend Georg Schweinfurth had announced the discovery of a new whale species (*Zeuglodon osiris*—now called *Basilosaurus*) from the rocks of the Fayoum. And in the same year that Stromer first visited the Fayoum, C. W. Andrews, from the British Museum (Natural History), announced finding there the teeth and jaw of one of the oldest-known elephants (*Palaeomastodon*).[18] Yet even though fossil hunters had by then been at work for almost a century, the fields of both geology and paleontology were still relatively young, and the conclusions of their practitioners about the age and origins of the Earth and of mankind, though widely discussed among scholars, were by no means universally embraced.

How rock layers were made—sedimentary layers, at least—had been explained in the mid-seventeenth century by Nicholas Steno, a Dane who lived for many years in Florence, working under the sponsorship of the Grand Duke of Tuscany.[19] There were three fundamental principles, said Steno, that govern the formation of sedimentary strata. His first principle, the *principle of superposition*, is perhaps the most obvious: In any sequence of undisturbed sedimentary rock strata, the oldest rocks are on the bottom, with the younger ones stacked successively atop them. His *principle of original horizontality* states that the tiny particles of material from which rock is composed settle through a fluid and arrange themselves in layers that solidify over time and are, initially at least, horizontal; any strata not horizontal have been disturbed by forces yet unknown. Finally, Steno's *principle of original lateral continuity* explains that sediment is deposited laterally in all directions until it runs up against some preexisting barrier or simply thins out at its margins.

These notions seem, in retrospect, virtually self-evident. But in the seventeenth century, they were revolutionary: To suggest that there were older or younger rocks was to suggest that the Earth was not created all at once, which contradicted everything believed at the time. That is to say, they contradicted the Bible, and in those days, contradicting the Bible was a dangerous business. After all, thanks to detailed calculations by the Irish bishop James Ussher in 1658 (later refined by Bishop Lightfoot), it was well known that the Almighty had created the Earth all at once, and at the remarkably civilized hour of nine A.M. on Monday, October 23, 4004 B.C.[20] Steno does not appear ever to have ventured an opinion—in public, at least—on the age of the rock layers he had so crisply described. He had explained *how* some rock is formed, but not *when*. Steno, after all, was a Catholic who would eventually become a bishop of the Church.

It was not until 1785 that someone endeavored to clarify how long it might take for such rocks to form. In that year, Edinburgh physician James Hutton published his *Theory of the Earth* and announced, no doubt to the horror of the clergy of the time, that he had found in his investigations of the rocks of his native Scotland "no vestige of a beginning, no prospect of an end"[21] to the history of the Earth. Hutton arrived at this conclusion from two blindingly simple observations: First, the rate at which sediments accumulate and rock erodes is very, very slow; and second, there is nothing to suggest that it has ever been otherwise. The next conclusion was as obvious as it was unsettling: The Earth as we know it must therefore have taken millions of years, not days, to create.

Meanwhile, at roughly the same time thoughtful gentlemen with a bit of spare time on their hands were figuring out how rocks were formed and beginning, tentatively, to speculate about when they were formed, others in England were struggling to make sense of peculiar objects farmers uncovered from time to time with their plows. The peculiar objects were stones. But they were stones in the shape, or with the imprint of recognizable objects: leaves, oyster shells, snails, sea urchins. They were puzzling because there was little in the received dogma of the late seventeenth and early eighteenth centuries that made it possible to explain what they were, since they were clearly none of these recognizable objects; they were rocks. For a while the only explanation was that these objects, called "figured stones," were placed by the Creator in a sort of heavenly version of showing off—that is, to prove He could do pretty much anything He wanted to, including embedding rocks that looked just like living things deep within the Earth. Later still, as people began to toy with the deeply troubling notion that the figured stones had once been living things but had somehow mineralized when they died, the question of how they got there remained. These stones, which were predominantly images of sea creatures, were often found far inland, even on mountaintops.

The ever nimble Church explained this problem away in a manner that did no violence to accepted dogma and actually seemed to enhance it: The mountaintop fossils were proof of Noah's Flood. The creatures had been swept to these locations by the Flood, became somehow petrified, and that was that. The fact that the Flood, like Creation itself, was an event measured in days, not the much longer period of time suggested by the rock layers, was a vexing complication. For now, at least, it seemed enough that the stones were miraculous and beautiful. For a while, even well into the mid-nineteenth century, these curiosities were all the rage, and rare was the manor-house drawing room that did not possess a glass-fronted case displaying the owner's much prized collection.

It took longer for another explanation—that the Earth must have much greater antiquity than ever before imagined—to gain currency. This idea had been suggested, albeit delicately, as early as the Renaissance by both Leonardo da Vinci and Girolamo Fracastoro of Verona,[22] but was long overshadowed by the more widely promoted biblical interpretation of Earth's creation.

It took a workingman, the son of a blacksmith and a virtually self-taught surveyor who himself embodied the concept of progress then emerging in rapidly industrializing England, to see the relationship between the principles of rock formation and the existence of figured stones in some rocks but not others, and then make that relationship visible and thus understandable. His name was William Smith. A denizen of coal-mine shafts and canal cuts, a man who traveled the length and breadth of England noting the rock formations he passed— not just their composition but also the angle and direction of their inclination—Smith was soon able to predict, to the astonishment of his friends, the precise rock composition and fossil content of places he had never even visited. The product of his genius, *The Delineation of the Strata of England and Wales with a Part of Scotland,* was the first ever geological map of an entire nation.[23]

Smith's map was published in London in 1815, only eighty-six years before Ernst Stromer entered Egypt for the first time. It would be another fifteen years after the map's publication before the English geologist (there now was such a profession) Charles Lyell would publish the first volume of his four-volume classic, *Principles of Geology,* which compiled the emerging knowledge of the day and added to geology's basic principles.[24] It was Lyell who proposed the *principle of cross-cutting relationships,* which stipulated that when one geologic feature cuts through another, the latter must necessarily be the older of the two—a concept that anyone who has wondered about the existence of veins of white quartz in dark masses of granite will find helpful, and that anyone searching for veins of gold or silver will find profitable. Lyell also added the *principle of inclusion,* which explained that whenever two rock formations are in contact with each other, the one that surrounds pieces of the other will be the younger of the two, having obviously flowed into or collected around the bits of the other.

For the great mass of people struggling for a living in the seventeenth and eighteenth centuries, the question of where rocks came from and what happened to them over time would have been considered a matter of interest only to wealthy amateurs. The question of where mankind itself came from, of how and when living things were created, was another matter altogether, however, one that cut much closer, as it were, to the bone.

So in the mid-eighteenth century, when people began to find fossils that looked like bones rather than shells—indeed, like the bones of nothing that currently existed on Earth—the issues of time and Creation became much more complicated. The trouble actually began a bit earlier, in 1677, when the curator of the Ashmolean Museum at Oxford University announced he was compiling a book on the natural history of Oxfordshire and was brought a large and puzzling fossil. It was cylindrical and had a broken end. Its center had been hollow but now was filled with solid sandstone, and it ended in two large rounded knobs. It looked like part of a petrified leg bone, but it was huge, much larger than the bone of an ox or an elephant or anything else known on Earth. In what must have seemed even to him a bit of a stretch, the curator, Robert Plot, reached back into mythology and proclaimed it the leg bone of a giant. He was not certain whether it was a male or a female giant. That puzzle was apparently resolved a century later, when another scientist concluded, rather surprisingly, that the fossil was an immense human scrotum.[25] Both, of course, would later be proved wrong.

Gradually, new and even more perplexing bone specimens were discovered. The Reverend William Buckland, dean of Christ Church at Oxford and a lecturer in geology, identified another group of fossilized bones and teeth as having belonged to a reptile, but one of unimaginable immensity. In 1824, after thinking about it for years (and no doubt praying for guidance), Buckland named the beast *Megalosaurus* ("big lizard"). By the mid-nineteenth century, several more fossils of what were believed to be giant reptiles had been discovered, described, and named. But while the naming of these puzzling creatures was one thing, explaining them was quite another.

The problem was basic: The Bible makes no mention of creatures being created and then becoming extinct. The Reverend George Young, who discovered a huge ichthyosaur (a large sea-living reptile, though not a dinosaur) near a village in Scotland, solved this problem, at least to his own satisfaction, by concluding in 1840 that scientists simply had not yet explored enough of the globe. When they did, he said, they would find living specimens of the beasts whose petrified bones were now being discovered.[26]

But science has a way of gaining velocity and sweeping all things before it. In England and elsewhere in the Western world, *dinosaurs* had captured the public's imagination. After the Great Exhibition of 1851 in London, Queen Victoria's husband, Prince Albert, had the Crystal Palace turned into a park and exhibition center for illustrating the history of the Earth. There were many exhibits, but the hit of the show was the reconstruction of an *Iguanodon*, another dinosaur. So large was the model that on the eve of the opening of the exhibit in 1854, the organizers held a celebratory dinner inside the hollow sculpted beast.

By now commoners and clergymen alike were forced to accept what scholars had been arguing for some time: both that the Earth was much older than previously imagined and that giant animals once roamed the planet but had somehow disappeared. But the biggest conundrum of all remained: the mechanism of that disappearance. Noah's Flood had long since been discarded as a credible explanation, but there was no theory with which to replace it.

As it turned out, another Englishman was working on the problem at that very moment. As the resident naturalist on a five-year mapping expedition around the world, the young man had been struck repeatedly by the fact that the fossils he found were similar, but not identical, to the living animals he captured or observed.[27] Though obviously related, they had changed. If life had been created, for all intents and purposes, at one time, then there was no explanation for the similarities and differences between extinct and living forms. There was only one possible explanation: Life had not been created all at once, it had evolved. Some years after his return to England, the naturalist Charles Darwin presented his comprehensive theory in the landmark study *The Origin of Species*. Extinction was explained by evolution.[28] Evolution was explained by the principle of natural selection. Darwin's book was published in 1859, only forty-two years before Stromer's first expedition to the Egyptian desert. Its effects were, and still are, far-reaching. But at the time it was published, one of its effects was to present to the young science of paleontology a purpose that lifted it out of dilettantism: to document through the ancient fossil record the history and process of evolution.

Ernst Freiherr Stromer von Reichenbach, the diminutive but determined man striding beside his camel on the morning of November 18, 1910, had little doubt that he and Richard Markgraf would be able to fill in important segments of that history, just as he had during his expeditions to the Fayoum Oasis a decade earlier.

As he made his way north from Cairo to Wadi el Natrun, Stromer was, in a sense, part of the vanguard of a new generation of paleontologists. Virtually everything that is known about paleontology in general and dinosaurs in particular has been learned in just the last two hundred years. But the first generation of paleontologists were principally desk-bound European gentleman scholars to whom bones, also European in origin, were delivered by amateur collectors and men who worked in quarries. In England, there were the aforementioned remarkable, and remarkably eccentric, Reverend William Buckland;[29] the physician Gideon Mantell; and the anatomist Richard Owen, who first coined the word "dinosaur" in 1842. There was Harry Govier Seeley, who first identified and named the two major groups of dinosaurs (Ornithischia and Saurischia, based on their differing pelvic structures), and several others who contributed significantly to the birth of the field. In France, there were Baron Georges Léopold Cuvier, the acknowledged father of the science of comparative anatomy (which made distinguishing among dinosaurs possible), and the melodiously named Jacques-Armand Eudes-Deslongchamps (whose collection, like Stromer's a century later, was destroyed in World War II). In Germany, there was the father of German vertebrate paleontology, Hermann von Meyer, who first identified the spectacular half-reptile, half-bird "missing link" *Archaeopteryx*.[30]

It was not until the late nineteenth century that paleontologists began to forsake their desks for prospecting in distant lands. Among the earliest field paleontologists were the prolific German Friedrich von Huene, who worked in Europe, South America, and South Africa; and his countryman Eberhard Fraas, who worked in the Fayoum Oasis and later uncovered the extraordinary dinosaur graveyard at Tendaguru, in what was then the German protectorate of East Africa (now Tanzania). But Fraas, weakened by amebic dysentery (which killed him a few years later), was forced to abandon the dig, and his work was carried on by Werner Janensch until World War I brought the excavations there to a halt.

But while all this was happening, a change was under way. The geographical locus of dinosaur hunting was shifting rapidly to the American West, and a Wild West it was, complete with shoot-outs among competing bone hunters. The first acknowledged American dinosaur paleontologist was the University of Pennsylvania anatomy professor Joseph Leidy. In 1868, at Philadelphia's Academy of Natural Sciences, Leidy reconstructed the first dinosaur skeleton to be displayed to the public anywhere in the world, a twenty-six-foot, bipedal herbivore found in New Jersey a decade earlier that he called *Hadrosaurus.* It was a University of Pennsylvania protégé of Leidy's, Edward Drinker Cope, and his rival, Yale University's Othniel Charles Marsh, who made dinosaurs a subject of widespread public fascination in the United States. Cope and Marsh had met in Europe and worked initially in the eastern United States. But as the railroads expanded into the West, construction crews began discovering immense ancient bones protruding from the treeless, arid ground and in the railway cuts they dug through hills. The two young paleontologists quickly shifted their attention westward. Working feverishly against each other, their diggers tore fossils out of the ground, hastily packed and shipped them east by boxcar, and raced to be the first to publish descriptions so as to secure the honor of naming the specimens. Wealthy as well as ambitious, they were not above establishing their own scientific journals in order to assure last-minute publication. It was part science and part circus. It was, in short, all-American.

In contrast to the grandstanding of the American bone warriors, European paleontologists worked quietly, even obscurely, producing detailed and often stunningly illustrated scientific monographs. In the first decade of the twentieth century, one of the quietest of these professionals was the student of the great geologist Karl Alfred von Zittel, Ernst

Stromer. It was in 1893, after studying medicine at the universities of Munich and Strasbourg for several years, that Stromer switched to geology and paleontology. At von Zittel's suggestion, he wrote his dissertation on the geology of the German protectorates in Africa and received his doctoral degree just before Christmas 1895. After further study in Munich and Berlin, Stromer was appointed conservator of the geology and mineralogy department of the Rijksmuseum in Leiden, Holland, in 1897, but he became ill and soon returned to Munich (he is almost invariably described as "frail," even though he lived to the ripe old age of eighty-two). Shortly thereafter, he was appointed lecturer in paleontology and geology at the University of Munich.[31]

To conduct the research for his dissertation on the geology of Germany's African colonies, Stromer traveled no farther than the university library. But after the turn of the century, the Sahara seems to have captured his imagination. Following his two winter expeditions to the Fayoum Oasis and Wadi el Natrun, Stromer spent the next decade producing nearly two dozen professional papers, most of them on his Egyptian fossil discoveries. In time he would become one of the most successful—if least known—field paleontologists of his era.

Stromer's 1910 journals of his days in Wadi el Natrun reveal a man thoroughly in his element. He worked throughout the day, apparently tirelessly, hiking for miles, climbing hills and escarpments, hammering out promising pieces of rock from outcrops in the valley. He recorded his activities in fastidious—one might say numbing—detail. Each journal entry is dated, his location given in precise coordinates, the times noted to the minute. His explanations of rock formations are precise, his descriptions of strata often colorful. They read like rough drafts of academic monographs, which is what they eventually would become. He confesses to weariness only at the end of these long days, complaining about the weight and poor fit of his specimen-filled knapsack.

As he scrambled over the valley slopes, Stromer yielded not an inch to the ruggedness of the desert climate or landscape in his manner of dress, apart from stout shoes. Pictures of him during this period show him dressed in Western attire—as, for that matter, do photographs of virtually all the European explorers of this age. From the poles to the Himalayas, there they stand in their expedition photos, looking for all the world as if they had just stepped out of a carriage in Piccadilly Circus or a tram in Berlin.

Stromer had visited Wadi el Natrun before and believed he might find there the kinds of rock layers that could contain the early mammal fossils he sought. But he would be disappointed. The weeks he and Markgraf spent in the wadi at the end of November 1910 turned out to be largely unsuccessful. Though he found no shortage of fossil fragments—shark's teeth, the smooth broken shells of ancient turtles, the jaw of the occasional prehistoric crocodile—he uncovered no mammalian remains.

By early December, Stromer had washed his hands of Wadi el Natrun and returned to Cairo to make arrangements for the second stage of the expedition. Markgraf remained behind to continue digging in the Natrun Valley upon Stromer's instructions, and not without success. When he finally returned to Cairo a week or so later, he was able to present his employer with the skull of a small monkey. Stromer was delighted. This was the fossil that would be named in Markgraf's honor, *Libypithecus markgrafi.*

Stromer may have been delighted by something else as well. While busying himself in Cairo with official and professional meetings during the first two weeks of December 1910, Stromer also had time to be entertained by a variety of friends and acquaintances, both German and British. Among the dinner parties he attended, often in the company of his friend Georg Schweinfurth, was one in the home of the Rennebaum family. In one aside in Stromer's journal, there is a hint of something to come: In a single sentence, he mentions having taken Fräulein Rennebaum—the Rennebaum's daughter Elizabeth—shopping for an afternoon. There is no mention in the journal of how Stromer felt about the fräulein, who would have been a very young woman. Perhaps he was only doing the Rennebaums a favor that day. Many years later, however,

Stromer chanced upon Elizabeth Rennebaum again, in Germany. This time he asked her to marry him, and in 1920 she accepted his offer. Though well into his fifties by then, he had three sons and a long marriage with Elizabeth. It was, however, a marriage that would be marked repeatedly by tragedy.

The second stage of Stromer's 1910 expedition took him to a location far up the Nile River in mid-December. It began pleasantly enough with a visit to some of the tombs and antiquities that had been excavated in Luxor. It is clear from his journals that Stromer enjoyed the excavations here immensely (though he noted that the addition of electric lighting had taken away some of the eerie romance from when the tombs were illuminated only by candlelight). Then Markgraf joined him, and the two traveled farther upriver, not far from Aswan, where they hired a few camels led by a Bedouin named Muhammad, who would also serve as Stromer's cook.

Markgraf, who once again was unwell, then returned to Cairo to make arrangements for the longer trip to the Western Desert. Stromer spent the next week and a half on his own, exploring rock strata exposed on the west-facing escarpments and side valleys of the Nile. He was looking for rock of an age and composition similar to what he had studied in Wadi el Natrun. But while he was often treated to sweeping views and stunning sunsets over the Nile from the slopes where he worked, these rocks proved no more fossil-rich than those in Natrun.

Disappointed, Stromer packed up on December 21, went to the nearest rail station, and boarded an overnight train for the 470-mile return trip to Cairo, arriving at seven-ten the following morning. After checking in to his hotel, Stromer went to the post office, where he found two letters that swung him from one emotional extreme to another. In the first letter, he learned that the second volume of a textbook he had written on paleozoology had been accepted for publication.[32] However, the second letter confirmed a fear that had been growing since he saw his collector a little over a week earlier in Luxor—Markgraf's illness had returned and he would not be able to undertake the most critical trip of all, the upcoming expedition to Bahariya Oasis.

With his limited grasp of Arabic and his complete unfamiliarity with the remote reaches of Egypt's Western Desert, Stromer suddenly felt abandoned and bereft. Now, two days before Christmas Eve, his entire expedition now seemed threatened, especially given the limited results of the trips to the Nile Valley and Wadi el Natrun.

It was a stunning setback.

THREE

UNEARTHING A LEGEND

"Whoa, stop . . . Go back!"

Reflexively, Josh Smith, in the passenger seat of a lurching Toyota Land Cruiser, flung out his left arm and smacked the right shoulder of his driver. It helped that his driver, Robert Giegengack, is a preternaturally imperturbable man of exceptionally good humor and an experienced hand at driving in the desert.

"Okay, okay," Giegengack shouted back amiably, swinging the vehicle in a wide arc to keep it from bogging down in the sand. "What the hell is it?"

"Back there. Something big right on the surface."

The Toyota had been rattling across the parched and rugged floor of the Bahariya Depression at roughly forty miles per hour, and Smith had been riding with his head out the passenger window, scanning the ground. The date was February 21, 1999. Smith, who is possessed of a pronounced and serious brow to begin with, was scowling as he squinted into the midmorning sun. He was worried. So far the day had not gone well. He had fewer than twenty-four hours to uncover a legend that had sat undisturbed, a mystery that had lain unexamined, for nearly a century. Smith, then a twenty-nine-year-old doctoral candidate in paleontology at the University of Pennsylvania, was looking for the lost dinosaurs of Egypt, the lost dinosaurs of Ernst Freiherr Stromer von Reichenbach. And he was getting nowhere.

It had begun, as is often the case with great notions, with beer. A year earlier, Smith had sat at the end of a long day in the office of a fellow Penn doctoral student and then-twenty-two-year-old dinosaur savant, Matt Lamanna, in the university's geology building: historic, red sandstone–sheathed Hayden Hall. They were working their way through a six-pack and musing idly about where on the globe they might begin a collaborative project.

"We started putting together a rough list of selection criteria," recalls Lamanna. "They had to be places that offered the prospect of success but also had not been worked to death. They had to have something about them that would permit us to make a real contribution to science, to what we know about the world. And, of course, they had to be cool places to go to.

"So, we had the usual locations where interesting new discoveries are being made—Patagonia, Mongolia, and a few other places—but what surprised us was that we each had this little known place, Bahariya, in Egypt, at the top of our lists. And we just sort of looked at each other."

In this regard, and in other ways that emerged in the months that followed, Smith and Lamanna revealed themselves to be different from many others in their chosen profession. Hans-Dieter Sues, vice president of collections and research at the Royal Ontario Museum in Toronto and one of the world's leading paleontologists, explains: "Generally, I would say that for the majority of paleontologists, there's a sort of herd mentality. You go places where other people are already working because it does have the distinct advantage that you're going to find something, too."

Sues adds another point that Josh Smith later found only too true: "Also, it's often a lot easier to convince funding agencies and sponsors to give you some research money for known places because they assume you'll bring something back.

"But there's a subset of our community," says Sues, "that thrives on being a bit more adventurous, going out on a limb, doing the kind of thing Ernst Stromer did—that is, saying to themselves, 'Let's not go to where everyone else is collecting fossils; let's go somewhere else!' They're the Indiana Joneses of our profession."

And when you examine a geological map of the world, Sues notes, and look for areas where rock from the age of the dinosaurs lies exposed on the surface, your eye cannot help but be drawn to North Africa. "There's a big stretch of Cretaceous rock that shows up there in various places. Rock from the Cretaceous Period—that is, rock anywhere between 66 and 144 million years old—is where many of the really rich dinosaur deposits have been found, like the badlands of the western United States and in Canada. In Cretaceous times, many of these areas were on the shorelines of great shallow seas. And that's exactly what you see when you look at a geological map of Egypt's Western Desert."

Josh Smith was drawn to Bahariya for another reason, though: "I've been fascinated with Egypt since I was in sixth grade; I distinctly remember doing a little paper on the pyramids and the pharaohs and being really captivated by that history and the desert in general."

Smith had other, less academic reasons for his interest in deserts. With grades that he admits were "less than spectacular," and a working-class family who could not easily afford to send him to college, Josh Smith made an obvious decision at the end of his senior year of high school: "For a six-year part-time commitment, the Army Reserve would pay for my [environmental geology] degree at the University of Massachusetts at Amherst. Pretty easy math." Like everyone else in the U.S. Army in 1990 and 1991—during the Gulf War—he had his eyes firmly focused on one particular desert.

But it was a decision that was more than purely financial. Smith had been, in his own words, "the quintessential skinny little science nerd" in high school. "I got beat up a lot," he recalls with a pain that is still evident. The army would be a way for him to toughen up and gain a certain level of self-esteem that he needed to acquire if he was going to get anywhere in his career—a career he already knew would involve physically challenging working conditions, not to mention the notoriously nasty infighting of the world of academia.

Soon after his advanced infantry training, and as he continued with his college studies, Smith began to seek out training that would make him an attractive candidate for selection by the army's Special Forces, the elite units that undertake some of the most difficult and hazardous missions in the military. Though his tour of duty expired before he acquired these skills, the legacy of Smith's military training goes far deeper than his characteristic crew cut and burly physique: He is an exceptionally capable, deeply determined young man who, once he receives an assignment or conceives a plan, drives forward relentlessly until it is achieved.

"All I know," jokes Matt Lamanna about his easygoing but intense friend, "is that I try not to make him angry; I'm sure he knows a hundred ways to kill people with just his hands."

A few weeks after Smith and Lamanna had talked about the sites they might one day like to explore for dinosaurs, Lamanna walked into Smith's office with a completed spreadsheet analyzing the criteria the two had listed for choosing an interesting field-expedition target. Sure enough, Bahariya was right at the top.

Says Smith, "There were several reasons to consider Bahariya, but three stand out for me, and Matt as well, I think. The first was no paleontologist had been there since Stromer. Second, very little is known about the terrestrial animals of Late Cretaceous Africa, beyond what Stromer and Markgraf found in Egypt, so there was an opportunity to make a real contribution to the science. Third was that everything Stromer found had been lost. Here was this pioneer, an explorer as much as a paleontologist, who identified a number of unique dinosaurs and a baroquely complex fauna in what now is a desert, and all his dinosaurs and many of his other finds were lost. All we have are his monographs. That seemed such a terrible injustice, not just to science but to this man himself, and we decided we wanted to try to recapture and rebuild his legacy. Someone should have done it long before now."

There are any number of good reasons why no one had returned to Bahariya since Ernst Stromer and Richard Markgraf first explored the oasis ninety years ago. For one thing, the Western Desert of Egypt is an exceptionally harsh environment in which to work. In the Bahariya Depression, which lies some 220 miles southwest of Cairo, temperatures in the summer can rise in excess of 130 degrees Fahrenheit. In the winter, the season when most work is done in the desert, temperatures can be bitterly cold at night and still viciously hot in the daytime. The threat of dehydration is constant in any season. The other constant is the wind and, borne upon it, sand and dust. Sandstorms are common year-round and at their worst in the winter and early spring.

War has been another barrier. Fierce and bloody battles raged across North Africa in both world wars, and the desert itself was a hotbed of espionage and intrigue.[1] For long stretches of the twentieth century, the Western Desert of Egypt was as inaccessible politically as it was physically. Even today much of the region is guarded by military patrols constantly on the watch for border incursions and internal insurgents. Every road into the Western Desert is marked by a checkpoint, and no foreigner ventures off established routes without a permit from the authorities in Cairo.

Finally, the plain fact is that until recently, those same authorities have seemed little interested in Egypt's paleontological history. What limited funds are available for geological endeavors in Egypt have been invested primarily in the exploration for mineral and energy resources.

Still, others have tried to reopen the strange case of Ernst Stromer and the Bahariya Oasis, but failed to get an expedition off the ground. Hans-Dieter Sues says, "Many of us have tried before. We did the proper thing: We applied in writing to the Egyptian authorities, requesting permission to enter the desert. When they do respond, the Egyptian authorities are very friendly and helpful. But mostly they do not respond, and you wait. And wait. You send a letter and nothing happens. You send a fax and nothing happens. This is largely a matter of culture; the Egyptians much prefer to conduct such negotiations in person, which is understandable and quite charming, but it makes it very hard to get an expedition organized and funded. Also, I think paleontology has just not been a priority for them."

Coincidence and chance often play critical, if rarely acknowledged, roles in the process of scientific discovery, and they would do so in the rediscovery of the lost dinosaurs of Egypt. It is not even clear whether Josh Smith understood at this early stage how difficult it could be to gain entry to Egypt's Western Desert, but two coincidences would come to his aid before he even knew he had a problem.

The first one sat, even as Smith was looking over Lamanna's spreadsheet, at a desk in an office exactly one story beneath and roughly six times larger than Smith's. The man seated at the desk was Robert Giegengack. Lanky, with penetrating eyes but a quick and genuine smile, he is known to his students simply as "Gieg." The man who, some months later, would be Smith's "driver," is chair of the University of Pennsylvania's Department of Earth and Environmental Science. A noted geologist, Giegengack had spent much of the last thirty years doing fieldwork on the climatic history of deserts. To Josh Smith's very good fortune, the places Gieg chose to pursue this research were the oases of the Western Desert of Egypt. Well known, well liked, and utterly trusted by officials of the Egyptian Geological Survey and Mining Authority (the successor to the original Egyptian Geological Survey), Giegengack would be the key to the door to the desert for his young student, a door that had been effectively closed to many paleontologists for nearly a century.

It is easy to understand why Giegengack is held in such esteem by the Egyptians: He is very fond of them and of Egypt. It is the place where, even more than the university campus where he spends much of each year, he feels most at home.

"Americans think of Egypt as a poor, dirty, hot, dusty place where it is hard to get around, hard to communicate, and hard to get what you need. All of this is true. But what is also true, what is far more important and what so many people, and not just Americans, fail to grasp, is that Egypt is one of the most civilized cultures on Earth. And it's not just superficially civilized. It's not a matter of there being technological marvels at every turn. There aren't many, and the fact is that in reality, they aren't important. What is important is the spiritual and cultural depth of a place and a people, and this is where Egypt really shines. It is a place where the people are rich despite having an absolute minimum of material possessions.

"There are a lot of deserts in the world. I could work in any one of them. I go to Egypt year after year because I love it. That is where my heart is."

As he sat in his office that day, Giegengack was wrestling with a problem that would become the second coincidence to make Smith's Bahariyan vision a reality. During the previous two seasons at the Dakhla Oasis in central Egypt, Giegengack had been assisted by one of his doctoral students, a student who had received multiple undergraduate honors at Harvard University before graduating magna cum laude in Earth and planetary sciences in 1996, a student whom Giegengack describes thus: "Without a doubt the most brilliant I have ever had. Period. A person of penetrating insight and absolute intellectual and personal integrity whose range of interests and depth of knowledge are simply astonishing."[2]

What worried Giegengack about the upcoming season in Dakhla was that, because of another professional commitment, he would have to leave for part of the season. With an assistant of this student's caliber, his departure should have been no problem. Giegengack could be perfectly confident the fieldwork would progress carefully and competently without him. But this assistant presented a problem. This assistant was a woman.

Jennifer Smith is a diminutive young lady, just under five foot three, with long dark hair, immense dark brown eyes, and a mouth that flashes with lightning speed from tightly pursed concentration to a broad and slightly feline smile that bears more than a trace resemblance to Julie Christie's. She also holds a black belt in karate.

For that, and two other reasons, Giegengack's concern was not Jennifer Smith's safety in Egypt. First, her determination is legendary. He explains with a story: "Once, on an expedition with Jennifer in the jungles of Belize on the Yucatán Peninsula, I saw a promising rock outcrop on the other side of what looked like a fairly shallow stream. I led the way across the stream, and Jen followed. Right away the water was up to my knees. I kept going, transferring the stuff in my pockets to a compartment in my knapsack. Then it rose above my waist, and I slid the knapsack off and held it over my head, which was a good thing, because in a few more moments the water was chest-deep. I'm embarrassed to say that I was most of the way across before I thought of Jennifer. I turned, and there she was behind me, completely submerged. All you could see were her two arms sticking above the water supporting her pack as she trudged through the water, hopping up from time to time to catch a breath. When she slogged out of the water on the other side, I was aghast and asked her why she hadn't said anything. She said simply, 'I was saving my breath.' "

The second reason Giegengack was unconcerned for her safety was the nature of Egypt itself. "Egypt is very safe; Cairo is a city of twelve million where violent crime is virtually unknown. Interpersonal violence is just beneath Egyptians, and women can walk safely through cities alone even at night."

Giegengack's problem wasn't criminal in nature, it was cultural. He knew that even with the proper entry permits, Jennifer Smith would never get past the checkpoint on the road from their lodgings to the research site in the desert. It would be inconceivable to the soldiers manning the gates that an unmarried woman, traveling alone, could do such a thing. It simply wasn't done. It would not be permitted.

Jennifer Smith, therefore, needed a male escort—better yet, a "husband." And quite by chance, in an office directly above him, Giegengack had a tailor-made candidate, a highly qualified field scientist who—coincidentally, since they were unrelated—had the same surname. It did not hurt that the two Smiths were already romantically involved and indeed would soon become engaged.

It is, unfortunately, in the nature of scientific research that specialists in one academic discipline—even closely related disciplines like geology and paleontology—often know very little about the history of each other's fields. So when Josh Smith agreed to join the expedition with the condition that, on the way to the Dakhla Oasis in central Egypt where Giegengack pursued his research, they stop for a night at the Bahariya Oasis, Giegengack was surprised.

"Frankly, after working in the Western Desert for thirty years," says Giegengack, "I had never even heard of Ernst Stromer or his dinosaurs. And anyway, I'm a geologist, not a paleontologist. If I ever did know anything about it, I had long since forgotten." Nonetheless, Gieg readily agreed to the stop.

It was, as he well knew, a bargain. Josh Smith was not a standard-issue graduate student. In addition to his military experience, he had already worked as a fossil preparator, received a number of academic awards, and published professional papers. Says Giegengack, "Even when he first arrived here, Josh thought, acted, and spoke like a professional. Since then he's produced professional papers prodigiously; in all the years I've been here, I've never seen a graduate student who's as focused as Josh is." Happily, he was also a skilled sedimentologist: just what Jen Smith and Giegengack needed in Dakhla.

"Still," admits Jen guiltily, "it wasn't an especially terrific deal for Josh; the price for spending a day or two in Bahariya was that he'd have to spend five or six weeks down in Dakhla working for me and Gieg.

"I'd done some paleontological work with Josh before, and I knew that even in fossil-rich locations, you can spend a number of eight- to ten-hour days hiking and prospecting and find only one or two little bones. Gieg and I figured Josh didn't stand a chance in hell of finding Stromer's dinosaurs in the middle of the desert in a day and a half. There was just no way."

The Toyota Land Cruiser crunched to a stop on a firm section of the desert floor and Josh Smith leaped out, walked a few paces, and bent down. At his feet were three sections of what once had been a very large bone. A dinosaur bone.

The sun was nearly at its zenith, and despite the fact that it was cold, the sunlight shimmered off the pebbly desert pavement. It had been a long, difficult morning, a morning of fruitless wandering and mounting frustration.

"Josh had some of Stromer's scientific monographs with him that day, but they weren't much help," recalls Jen, who was riding in the backseat. "The papers described geologic strata and the details of the bones he found. Like most scientists, he didn't waste words on the land-scape or directions, though he did mention an 'isolated conical hill' called Gebel³ el Dist. If we could find the general area where Stromer had worked ninety years ago, we could probably identify the site from his description of the rock strata there. But first we had to find the hill."

As it turned out, a friend of Giegengack's, Bahay Issawi, the former head of the Egyptian Geological Survey and Mining Authority, had photocopied a paper about one of Stromer's finds that included survey coordinates for Gebel el Dist.

"If you're looking for invertebrate fossils—little critters without a spine—or for ancient plants," says Josh Smith, "you work a specific, often very small site, sometimes for years. But with large vertebrate fos-sils, that's not as important, because the productive horizons are spread over a much broader area. If we got into the same general suite of rocks where Stromer had worked, we would have just as high—or low— a probability of finding bones as he had. We knew he'd found his di-nosaurs mostly near Gebel el Dist, so that's what we went looking for that first morning. We punched the coordinates that were in the re-search paper into our global positioning system [GPS] and headed northeast across the desert from the town of Bawiti, where we had spent the night. We had the technology; it should have been easy."

It wasn't. The three scientists bumped across a landscape marked by not a shred of greenery. For as far as the eye could see, there was only an uneven, rather lumpy expanse of rock-strewn grayish-tan earth and sand. Eventually they approached two long, low ridges, each a bit more than a half mile in length separated by a saddle. Giegengack drove the Toyota around one end and came up along the back of the ridges. Here the driving became dramatically more difficult. The ridges were capped with a hard limestone layer that had eroded away along the southeast-facing side. The ground was strewn with enormous chunks of limestone and, in their lee, drifts of soft sand. Several times they sank up to the Toyota's axles and had to spend as much as an hour digging out.

"Okay," Jennifer suddenly announced as they lurched along the slope, "we're almost on top of the coordinates."

Josh and Giegengack looked at each other blankly. There was nothing here that even remotely resembled the features they were looking for. They were simply running along the dip slope of the linear ridge and there was nothing isolated or conical about it. There was one spot where the ridge had slumped slightly and eroded, but Josh couldn't make himself describe the feature as either isolated or cone-shaped. They stopped, got out, and took a look around for a while anyway, wandering across the wind- and sun-blasted desert floor but seeing nothing.

"I figured, you know, forget it," remembers Jennifer, "we have obviously screwed this up somehow. And it's not like you can ask anyone for directions. First of all, there's no one out there; we were in the middle of nowhere. Second, place-names in the desert are mutable; one person will call a hill one name and another will call it something else."

"We'd spent most of the morning getting there," Josh recalls, "and we hadn't seen much at all. The day was moving on, it was cold, the wind was kicking up, and we were tired, hungry, and frustrated. At some point Jen looked across the depression to the west and said, 'Why don't we try over there somewhere?' and I just shrugged and said, 'Okay, fine, whatever. This place is getting us nowhere anyway.'"

But even as Giegengack swung the Land Cruiser around and hurtled off toward the northwest, Josh Smith was still searching the ground, his head out the window.

"I have never known a student who is as doggedly determined as Josh Smith," Giegengack says. "He's a terrifically hard worker. I joke that he is from the 'brute force' school of science; he will beat a problem to death, consider every possible aspect of it, to reach a solution. He will keep at a thing long after other people have decided it's a dead end. That's the downside. The upside is that he often succeeds."

And succeed he did. Now, as he squatted on his haunches above a group of very large bone fragments, the largest of which measured eighteen inches long and seven inches across, the tension that had been in Josh Smith's face disappeared. From the largest of these fragments, he knew instantly that he was looking at a portion of the leg bone of a sauropod, an immense, long-necked, long-tailed, plant-eating dinosaur.

"Sweet," he said quietly.

Since the day in 1842 when England's Sir Richard Owen first created the term "dinosaur" for the strange creatures whose petrified bones were being found, hundreds of different dinosaur species have been discovered. Gradually, with the bones as guides, scientists have been able to piece together a reasonably understandable if still fragmentary family tree, based upon the shared characteristics of different groups of dinosaurs.

The stocky stem of this tree is the broad category Vertebrata, animals with backbones. Vertebrates are divided between fish and those with limbs, called tetrapods. These in turn are divided between amphibians, which return to the water to reproduce; and amniotes, creatures that effectively carry the water with them, within either an egg or a womb. Amniotes diverged into two lineages: synapsids, which eventually led to mammals, and reptiles. The reptile lineage continued to evolve and branch for millions of years before the arrival of one distinctive and highly successful subgroup: dinosaurs.[4]

Dinosaurs distinguished themselves from all other creatures on the Earth roughly 230 million years ago by standing upright on two strong hind legs, their center of gravity in front of their hip junction, counterbalanced by a long and heavy tail. The dinosaurs split into two groups, ornithischians and saurischians. Both of their names are misleading. The name ornithischian means "bird-hipped," while saurischian means "lizard-hipped." But birds do not come from the ornithischian line; nor are lizards saurischians. The confusion stems simply from the shape and orientation of one bone in the pelvis, the pubis. In ornithischians, the pubis points backward beneath the hip (as it does in modern birds); in most saurischians, it points forward (as it does in modern lizards). No one knows precisely why.

Over time, ornithischians evolved into three broad groups. The first was Thyreophora, which produced the heavily armored dinosaurs, the best known of which are probably the stegosaurs, which had spikes and plates atop their backs and tails. The second was Marginocephalia, which includes forms that had thick, bony skulls and horns, like the amazing *Triceratops*. The third was Ornithopoda, which includes mostly bipedal beaked and duck-billed dinosaurs. All three groups were herbivores, or plant eaters.

But it was the second broad category of dinosaurs, the saurischians, that would come to dominate what today is Bahariya 95 million years ago (though ornithischians would predominate in many other areas of the Northern Hemisphere during the Cretaceous Period). Within this category there were two primary groups: the sauropods and the theropods. The former were, like the ornithischians, herbivores. Sauropods had small heads, long necks, and equally long tails. They grew to such enormous size that they had to walk on all fours, on elephantlike legs nearly as tall and thick as building columns. Emerging roughly 210 million years ago and evolving over tens of millions of years, they ranged widely in size while maintaining the same basic body characteristics. The longest sauropods (diplodocids) reached over a hundred feet, and the heaviest (titanosaurs) may have weighed as much as a hundred tons.

Big as the sauropods were, it is the theropods that get all the attention. Most were fearsome carnivores. Smaller and faster than sauropods, they prowled the landscape on two powerful hind legs that ended in feet with three functional toes. Their forelegs were generally much smaller by comparison. Theropods appear to have emerged roughly 225 million years ago, and then diversified. They were abundant throughout the Mesozoic Era and ranged in size from the forty-three-foot-long *Giganotosaurus*[5] to the pigeon-sized *Microraptor*. Theropods are the only kind of dinosaur that survives today—amazingly, as birds.

As he bent over the sauropod bones in Bahariya that day in February 1999, Josh Smith was relieved but not particularly excited. For one thing, Smith is a theropod guy. He already knew from Stromer's work that there were sauropods in Bahariya; finding the remains of one was not all that surprising. More important from a paleontological perspective, the pieces of bone were simply sitting on the surface, not protruding from the ground, which would have suggested that more lay beneath. As it was, the bone was what paleontologists call "float"—fragments left behind by the elements. Given this, Josh and Jen and Giegengack concluded that these fossils were of no special scientific significance. After marking the coordinates of the bone and photographing the fragments, they left. It would be more than a year before they returned to this site to discover, to their very great relief and to the amazement of much of the rest of the world, that their initial conclusion had been wildly incorrect.

Now, however, Josh, Jennifer, and Gieg drove west across the floor of the Bahariya Depression into the lowering sun. The afternoon was slipping away, and shadows had begun to lengthen in the lee of the hillocks they passed. After another hour and a half of driving around the desert floor, they topped a rise and saw ahead a distinctive landform silhouetted in black against the deepening orange of the early winter desert sunset. They recognized it immediately.

It was Stromer's isolated cone-shaped hill. It was Gebel el Dist.

The next morning, Josh was buzzing with nervous energy. With Giegengack, he dragged the slow-starting Jennifer out to the Toyota, shoving a thermos of black coffee in her direction, and headed them out for the northwestern desert. On this morning, partly because they now knew what they were looking for, they couldn't miss Gebel el Dist, but it still took more than an hour of rugged driving to get there.

"The moment we reached el Dist," Josh remembers, "things just got stupid. I mean, there was bone all over the place, all around the base of that hill. Lots of it. I had just returned with Matt Lamanna from an expedition to a very rich dinosaur fossil deposit in Argentina, but there was much more bone on the ground here in Egypt. The sheer quantity of float was astonishing."

While Jennifer prowled along the crest of a ridge, Josh and Gieg walked around the base of a small hill a short distance from el Dist.

"And there, right at my feet, I found another very large bone," says Josh, "a hollow, shaftlike bone, almost like a stovepipe, that could only have come from a very large carnivorous dinosaur, a theropod. The sauropod the day before was interesting, but this was the bone that started everything rolling. And there were more bones surrounding it. It was really exciting."

But Josh's Bahariya hourglass had run out. It was time to turn south to Dakhla to begin the real work of Giegengack and Jennifer's expedition. Josh didn't care; he'd found what he came for: Stromer's dinosaur graveyard. Soon, however, he would face months of the taxing pick-and-shovel work of raising funds for an expedition to Bahariya—a skill for which there is no course in graduate school and in which luck is an essential ingredient.

On March 10, 1999, the day after his return from Egypt, Josh Smith thundered up the stairs of Hayden Hall and into Matt Lamanna's office.

"I heard him before I saw him," Lamanna recalls. "He has this way of stomping up the stairs. But when he came around the corner, I hardly recognized him. He had this long messy hair and a completely overgrown red beard. He looked like a wild man. He also had this huge smile on his face. He smacked down a pile of photographs and just said, 'Dude, check it out.'

"I looked at the pictures and could hardly believe my eyes. Here was this barren desert pavement just strewn with bone. I'd never seen anything like it."

Smith remembers, "I showed him one picture, and his eyes got as big as dinner plates and he said, 'Is that a theropod?' I said, 'Yeah, I'm pretty sure it is, but what kind?' And he looked at some more and just said, 'I'm not sure yet; I'll let you know. But I can tell you this: Whatever it is, it is very seriously big.' "

Smith made one other important stop right after he returned from Egypt, at the office of his faculty adviser, Peter Dodson. A professor of anatomy, geology, and dinosaur paleontology wrapped up in one genial package, Dodson is a sort of one-man Earth sciences department. He has worked on dinosaur expeditions in Madagascar, China, and North America. Along the way he has written or edited a number of books on dinosaurs, including the award-winning reference book *The Dinosauria*, as well as books for children and scores of professional papers. But Dodson's most important products, as far as he is concerned, at least, are his graduate students. He has sent off from the University of Pennsylvania several of the most successful young paleontologists working in the world today.

Of Dodson, Smith recalls, "When I started looking around at grad schools and talking to professionals in the field of paleontology, Peter Dodson's name and Penn kept showing up. And I decided, why look anywhere else?" Says Lamanna: "I came from a very small town and really didn't want to be in a place as big as Philadelphia, but Peter took me under his wing. As an adviser, he's terrific. He doesn't tell you what to do, he encourages you to seek for yourself, and then he supports you all the way. It really builds up your confidence; I think that's why he's had so many successful students."

Smith showed Dodson the photographs, reviewed the strange story of Ernst Stromer's lost dinosaurs, and asked his adviser what he thought. "Well," said Dodson, "I think we'd better buy some airplane tickets."

In the days that followed, however, Josh had very little time to enjoy the enthusiasm of his colleagues and his adviser. He and Jennifer already were busy preparing for a second seven-week expedition, this time into the Arctic. Amid the preparations, Josh met a friend of his, an evolutionary biology doctoral student named Scott Winters, for a beer at the New Deck Tavern on the edge of the Penn campus.

"We were sitting in a booth in the back, just talking about the Egypt trip, and I showed him my pictures and he just said, 'Listen, I have an idea; can I show these to some friends of mine?'"

Even as he stepped out the door of the New Deck, Winters was punching numbers into his cell phone. While a student, Winters was also a partner in a small filmmaking company, Last Word Productions, and he immediately began looking for a larger firm that might be interested in underwriting an expedition to Bahariya in order to make a documentary. By the time Josh and Jennifer returned from the Arctic, there was an e-mail waiting from Winters. He had pitched the idea to a film company in Los Angeles called MPH Entertainment, which had asked for a proposal. Josh obliged, and shortly after MPH received his proposal, they wrote to say they were interested.

Meanwhile, Josh Smith was preparing to present a talk at the annual meeting of the Society of Vertebrate Paleontology, the principal professional organization in the field, in Denver that October. The presentation he made was like nothing the society had heard before.

Hans-Dieter Sues, the society's president, remembers it vividly: "I shouldn't say this, perhaps, but often at these meetings you just sit through talk after talk and you wonder, Jesus, why is anyone wasting their time studying that stuff, you know? Because of course they're not doing what you're doing, so you're not that interested. And people are constantly getting up and going to other sessions they think will be more interesting. But word had gotten around that Josh's talk was the one to hear, and people were just flooding into the hall. And once he started, nobody moved. He didn't begin by talking about dinosaurs at all."

"I just decided to do something different," explains Smith. "I said, 'Well, this morning we've been hearing all sorts of things about theropod phylogeny. Now we're going to talk about war.'"

"My first thought," says Sues, "was 'What is this, a Veterans of Foreign Wars meeting?' But then Josh started talking about his own military experience, and about the incredibly destructive effects of war on all of society. Then he told the story of Stromer's losses in World War II, his rediscovery of the place in the Egyptian desert where Stromer had worked, the amazing amount of bone still there, and the expedition he was planning. The audience response was amazing. I remember thinking, 'Wow.'"

Sitting in that audience, Mark Hufnail was paying very close attention both to Smith and to the response he was getting. Mark is the "H" in MPH Entertainment, the company that was considering underwriting the Bahariya expedition to film a documentary on the dig. But at this point, MPH had yet to commit to the project financially. The company filmed the presentation, interviewed Josh, Jennifer, Matt, and Peter on-camera, and, shortly thereafter, began shopping the film proposal to several major cable-television companies. No one bit.

"Hell," says Josh now, laughing, "I wouldn't have, either. To begin with, we were only grad students. What's more, we were proposing to go someplace unknown, or at least long forgotten. And to top it off, we hadn't actually found anything yet."

To their credit, and to Josh's everlasting gratitude, MPH didn't give up. But Smith was running out of time.

"As Christmas of 1999 approached, we still didn't know whether we were going back to Bahariya in January or not. I'd already assembled an expedition team and we had arranged things like passports, inoculations, preliminary permits, and excavation supplies. But we had no idea whether it was on; it was that up in the air, that shoestring an operation."

It was a situation Ernst Stromer and many other early explorers would have understood only too well. In the years before World War I, most scientific expeditions were funded not by governments but by wealthy patrons (Stromer, for example, occasionally received support from the Krupps von Bohlen und Halbach family of industrialists, who had an interest in paleontology), as well as by small grants from universities or scientific societies and, not incidentally, by the explorer's own personal resources.

Not surprisingly and almost by necessity, given these financial constraints, most explorers and field scientists in the early days of the twentieth century had, like Stromer, aristocratic backgrounds and independent means. But while Stromer was an aristocrat by birth, the evidence suggests that he was by no means wealthy. The total budget for Stromer's 1910–11 expedition, including shipping, was eighteen thousand German marks, a paltry sum, which would certainly help explain his anxiety about the sudden appearance of Dr. Leuchs's wife. In an era when African desert expeditions typically involved immense caravans and dozens, even hundreds, of native laborers, Stromer's own expeditions were astonishingly modest—a few camels and sometimes just one Egyptian helper who served as camel driver, cook, and guide. It was something about which he was especially proud, and it is one of many ways in which Stromer revealed his essential nature: efficient, exacting, careful with every detail, and a bit of a tightwad.

Neither Josh Smith nor his colleagues had the luxury of financing the Bahariya expedition themselves. While they attended an expensive Ivy League university, they were, to a man and woman, from blue-collar backgrounds and supported largely by their own labors and occasional fellowships.

"Peter Dodson secured a $5,000 university grant as well as an additional grant of $10,000 from a local philanthropist, the late Emilie de Hellebranth, and the project received a $1,000 donation from the Delaware Valley Paleontology Society," says Smith. "All together, it came to only $16,000. We needed at least $60,000.[6] We were a very long way from being able to afford to go."

Eventually, in December, Hufnail and his two partners, Jim Milio and Melissa Jo Peltier, met to decide what to do next. They knew funding and filming the expedition themselves would put the financial security of their own firm at risk. They voted. The vote was two to one in favor. As Milio put it later, "This is a democracy, and the majority rules." What Milio does not generally disclose is that he placed a second mortgage on his own home to cover the cost of funding and filming the expedition. Eventually, a major new force in science-inspired entertainment—Cosmos Studios, founded by the late Carl Sagan's wife and colleague, Ann Druyan—would step in to back the project and reimburse both Milio and MPH. Indeed, Cosmos gave MPH more than they had asked for; excellence was Cosmos's objective, and they were willing to pay for it.

In the frantic weeks that followed, Smith scrambled to complete the travel arrangements for all the participants in what henceforth would be known as the Bahariya Dinosaur Project. Finally, on January 10, 2000, the team boarded a plane in Philadelphia bound for London and then Cairo.

The search for the lost dinosaurs of Egypt had begun.

FOUR

DRAGOMEN, FOSSILS, AND FLEAS

Almost exactly ninety years earlier, on the eve of the second decade of the twentieth century, Ernst Stromer was in a similar state of frantic preparation.

In the week since Christmas, he had struggled with the unanticipated calamity of his partner's illness, and had swung, pendulum-like, between fear and determination, frustration and action. He knew his own weaknesses only too well. His Arabic was rudimentary. His knowledge of the Western Desert beyond the Fayoum Oasis was next to nil, limited to what he had gleaned from the report written on Bahariya by British surveyors—a report he now studied closely. He was not accustomed to traveling without a guide and cook (indeed, his good friend Schweinfurth had criticized him for carrying his own knapsack). What's more, he worried over the limits of his own physical health. Writing in his diary the day after Christmas, he observed, "I am noticeably tired as a reaction to all the days I have been traveling. I am, after all, too frail for such expeditions; I substitute tenacity and flexibility for the strength and pluck that I lack."[1]

For several days, Stromer sought a replacement for Markgraf, some other fossil collector conversant in Arabic and in the ways of the desert to whom he could entrust the details of organizing the trip, managing the recruitment of laborers and camels, guiding him to Bahariya, and assisting with exploration. Eventually, a friend of his in Cairo recommended a candidate, a gentleman named Hartmann. Stromer spent two days trying to locate him, writing to his postal address and traveling to restaurants and cafés the man was known to frequent. "One loses so much time here just trying to find someone," he wrote, "because no one knows the exact addresses and house numbers anywhere and address books are available only in hotels—but not in mine."

Stromer finally reached the collector, and Herr Hartmann presented himself at Stromer's hotel one afternoon a few days before the end of the year. "The somewhat gray-bearded Hartmann arrives," Stromer wrote with evident disgust, "in a dirty suit, spreading an indescribable odor. He appears to be down and out, denies having any experience in the desert and declares that he does not want to occupy himself with fossil gathering or chiseling anything out of rock and is uninterested in learning how to do it." Hartmann, in fact, suggested caustically that what Stromer wanted was an Italian stonemason, not a man of his more elevated scientific talents. Stromer dismissed him, not without relief, noting that the man's odor "contaminates my room long after his leaving."

Stromer found deliverance, of a sort, in a dragoman (a guide and translator) recommended by the German consulate, a Mr. Mohammad Hasranîn el Hitu, who worked as a guide at the luxurious, and ultimately infamous, Sheppards Hotel—a hotel designed, in fact, by Elizabeth Rennebaum's father, Johann Adam. Stromer describes the man as "young and dressed elegantly" in the uniform of the hotel, and they spent an hour discussing the expedition proposition and haggling about prices. Eventually, although not amicably, they arrived at an agreement.

During the following days, Stromer purchased "a pair of cowhide shoes with thick rope soles recommended to me for marching in the desert sand," canned and bottled provisions from Giorgioni's market, and several bottles of Palestinian wine. A friend visited and gave Stromer a hundred pieces of a rock-hard, half-moon-shaped type of bread that could be reconstituted in water and baked in the ashes of a fire. And he collected old newspapers with which to wrap any fossils he might find on the expedition. In addition, he obtained another water crate to augment the three he had been traveling with—this one, he noted, with a lock that would prohibit his workers from draining it on their own.

In the meantime, Stromer crisscrossed Cairo trying, unsuccessfully and with mounting frustration, to secure the permits he would require from British, French, and Egyptian authorities to travel in the Western Desert. It was a process so labyrinthine and inefficient as to drive his orderly German mind to the limit of its patience. Despite multiple visits to assorted offices, telephone calls, telegrams, letters, and still more visits, he could not get the authorities to act expeditiously, even as the day and hour of his departure loomed. In the end, he was forced to ask British authorities to telegram the mamur, or chief police officer, in Bahariya and announce his forthcoming arrival. Even then, he did not receive the appropriate papers until a messenger delivered them to his hotel in Fayoum after he had already left Cairo.

Finally, on the afternoon of New Year's Eve 1910, Stromer boarded a train to Medinet el Fayoum, the principal town in the Fayoum Oasis. Upon his arrival late in the afternoon, he was met by Mohammad Maslim, a servant he had employed during his earlier expeditions to this oasis but who now, nearly a decade later, had aged so much Stromer barely recognized him. Maslim was so pleased to see him and so eager to serve that Stromer hired him, despite his misgivings, for the trip.

After a bedbug-plagued night in a cheap hotel, Stromer spent most of the next day waiting for his dragoman, el Hitu, to arrive from Cairo by train. The camels would not arrive until the following day and when they did, to Stromer's irritation, one of the four proved to be a juvenile. What is more, though Stromer had contracted with el Hitu for four drivers, only two arrived. In the end, this would prove to be a boon, for Stromer found a third camel driver in Fayoum, a Bedouin named Gumaa, who knew the desert caravan routes far better than Stromer's alleged "guide," el Hitu. Indeed, el Hitu had already become a thorn in Stromer's side. He mistrusted the dragoman. Moreover he resented el

Hitu's tendency, in Stromer's eyes at least, to attempt to rise above "his station." Here Stromer revealed his least likable trait: He was an unreconstructed colonialist snob. His journal makes it clear that he regarded the Egyptians he hired as inferior beings—ignorant, dirty, greedy, devious, even dangerous. The great irony, one that no doubt would have been lost on a Victorian-era aristocrat like Stromer, was that he depended upon these people utterly—to manage the camels, to guide him through the desert and keep him safe, to cook for him, and to carry burdens.

Finally, on the morning of January 3, after a day of frustrating efforts to get the travel permits signed by the appropriate authorities in the Fayoum, Stromer, el Hitu, and Maslim boarded a train for Ghauraq, the end of the line, at the southern edge of the oasis. As the steam train chuffed across the oasis floor, Stromer marveled at the lush agricultural landscape through which they passed: "Everything is green, the grain is already a meter high and I can see some of the heads are already blooming; the sugar cane is also ripe and the maize has already been harvested and the stalks are layered on the ground, drying." It was the last greenery Stromer would see for many days. That afternoon they set up camp in Ghauraq near a water source where they could fill the four water crates he had brought. They waited there until the camels caught up with them. Stromer had brought a simple canvas wall tent, supported by two vertical poles and held down by guy ropes staked in the sand. Maslim assumed his accustomed place just outside the entry to the tent and began preparing a simple dinner of chicken and rice. The expedition to Bahariya was finally under way.

By noon the next day, they were deep in the desert, making for Ain el Rayyan, the spring at the head of the Rayyan Valley, southwest of Fayoum. But because of el Hitu's dawdling to graze and rest his camels, Stromer's little party failed to reach the spring that first day. It was a pattern that would repeat itself for the rest of the trip. Stromer had suspected that el Hitu had scrimped on purchasing fodder for the camels, and now he had proof. At every opportunity, the dragoman led the caravan off on long detours to known or suspected grazing areas, which, given their rarity in the desert, slowed the caravan's progress dramatically. There was little Stromer could do about it but fume. He was utterly at el Hitu's mercy. Markgraf, he knew, never would have put up with it.

The high desert plateau between the Fayoum and Bahariya oases is a landscape only a geologist could love. There are vast, undulating expanses of wind-eroded sandstone and limestone with shifting sand between the hillocks. There are crumbling outcrops of limestone here and there and, on occasion, mesa promontories in the distance. The sunbaked rock is so bright, even in winter, that Stromer often wore his yellow prescription sunglasses. It was also cold. Stromer, who was inadequately dressed, often walked beside the camels rather than riding, simply to generate enough heat to stay warm.

He wrote in his journal constantly, painstakingly recording even the slightest changes in the geology around him and in the distance, noting hard white limestones, greenish-white sandstones, crusts of greenish gypsum, pavements of light yellow to ocher sandstones, outcrops of quartzite, reddish-brown claystone, and even limestone tinted lilac. There were oddities as well: long bands of dunes, always oriented by the prevailing winds in the same northwest/southeast direction, fields of weathered limestone accretions, called "ball stones," scatterings of broken and petrified wood Stromer called "wood pebbles," and, numerous beyond belief, literally billions of fossilized oyster shells and *Nummulites*, single-celled foraminifera as large as thumbnails, along with scattered bits of ancient coral.

All of these deposits and rock formations spoke of a world that once lay beneath a warm, shallow sea—a sea that advanced and retreated many times over many millions of years, laying down alternating bands of shell-packed limestone and fine-grained sandstone. Stromer knew all this beforehand, of course, from the detailed reports published by the Egyptian Geological Survey, but he was mesmerized nonetheless. As the camels and their drivers plodded across the desert floor, Stromer often scurried into the distance with his rock hammer to examine closely some interesting outcrop, or simply dropped to his knees to marvel at the astounding density of the fossil shells beneath his feet.

Almost obsessively, as if committing the landscape around him to print before it blew away forever—which, in a sense, he was—Stromer described what he saw, his detailed journal entries seldom more than a few minutes apart. At night, his sleep was broken constantly by a plague of biting fleas, and early each morning he continued to write the record of the desperately bleak world through which his little band traveled.

Finally, on the morning of January 11, after more than a week of marching, Stromer noticed that the camel drivers were pushing their charges at a faster clip. He realized they must be approaching their destination, the Bahariya Depression. As they traversed the crest of another band of dunes more than twenty feet high, he could see the valley at last. Soon he was at the edge of the precipice. He reported in his journal, "we made a sandy descent steeply down into the Bahariya valley."

As his camels picked their way carefully down this cleft in the rim of the great Bahariya Depression, the expanse of landscape that opened before and below Stromer was balm to his eyes: "surprisingly beautiful after the monotony up to this point." His predecessors, the British surveyors John Ball and Hugh Beadnell, noted in their 1903 report, which Stromer was carrying, that "A fine view of the depression is obtainable from the top of the escarpment, a broad low-lying expanse, bounded by steep escarpments or walls, stretching away to the south, its monotony relieved by several large flat-topped hill-masses, near which, in the lowest portions of the floor, dark areas, the cultivated lands and palm groves can be distinguished."[2]

They also described its form: "In plan the oasis is of highly irregular outline, more particularly on its western side; but the general shape of the excavation is that of a large oval, with its major axis running northeast and south-west, and with a narrow blunt pointed extension at each end."[3] Though it is smaller than the other oases in Egypt's Western Desert (and was known for centuries as al-Waha al-Saghira, or "Little Oasis"[4]), the Bahariya Depression is nonetheless vast. Distinct from the other oases in the Western Desert, it is completely enclosed by a deeply scalloped limestone-capped escarpment; its floor measures nearly sixty miles long and more than twenty-five miles wide at its broadest point.

Oases in general bear little resemblance to the paradises of film or fiction. One early oasis explorer cautioned that the impression of lushness "is in great measure fictitious; it has chiefly its origin in the relief afforded to the mind, wearied by the monotony and dreariness of the surrounding wastes."[5] Bahariya is no exception. Although the total area of the oasis is roughly 1,100 square miles, only a tiny percentage of this area has water and is therefore cultivated. Everything else in the depression is as dry and desolate as the rest of the Western Desert, only lower; the average depth of the floor of Bahariya Oasis is 295 feet below the level of the surrounding plateau.

As Stromer, the dragoman el Hitu and his camel drivers, the servant and cook Mohammad Maslim, and the Bedouin Gumaa descended toward the oasis floor, they were effectively walking several tens of millions of years back through time. In the century after the publication of James Hutton's *Theory of the Earth* in 1785, researchers in the infant science of geology were beginning to piece together the Earth's history, rock by rock, fossil by fossil. The result was a sort of vertical map that came to be called the Geologic Time Scale. In a sense, this map was to become the visual embodiment of the principles established by the earliest geologists—for example, that younger sedimentary rocks are superimposed upon older ones and that, when deposited, they spread laterally, and predictably, in all directions to form distinct layers.

But in its early incarnations, the time scale's name was misleading, for it mapped rocks only in their relationship with one another in space, not in time as we now understand it to be. It could not, because as recently as the early years of the twentieth century, no one had the slightest idea precisely how old the Earth was. It was not until scientists understood the process of radioactive decay and the concept of the predictable half-life of certain elements that they were able to calculate the Earth's true age.

Scientists now estimate that the Earth is some 4.6 billion years old. This is an almost unfathomable period of time. Human civilization—

that is, the point at which our ancestors changed from wandering hunters and gatherers to settled farmers—is perhaps ten thousand years old, almost invisible on a time scale measured in billions of years. Dinosaurs did not even appear on Earth until 230 million years ago, after more than 95 percent of the Earth's total history already had passed, and 165 million years later they were gone.

The Geological Time Scale today divides the Earth's entire history into four immense blocks of time called eons. From the oldest to the most recent, they are the Hadean, Archean, Proterozoic, and Phanerozoic.

For roughly the first billion of the Earth's 4.6 billion years—or all of the wonderfully named Hadean Eon and half of the Archean Eon that followed it—there appears to have been no life of even the most primitive kind on the planet. The first single-celled organisms, bacteria and algae, appeared toward the end of the first billion years and spread for the next three billion as indeed they continue to do today. The great strength of some of these organisms was their ability to convert the gases in the early atmosphere and seas (which were themselves changing as a result of photochemical reactions) into food and oxygen. They did this for a very long time before they were succeeded by single-celled organisms with a defined nucleus and, even later—toward the end of the Proterozoic Eon—by more complex but still tiny multicelled lifeforms. The nearly four-billion-year span of time covered by the Hadean, Archean, and Proterozoic eons is known now as Precambrian—which is to say, it came before ("pre-") the first period ("Cambrian") of the most recent eon, the Phanerozoic.

It was in the Phanerozoic ("phanero," evident; "zoic," animal life) Eon that organisms with hard parts—bones and shells—first appeared in any abundance. The Phanerozoic Eon is divided into three vast eras. The Paleozoic ("ancient life") is the earliest, followed by the Mesozoic ("middle life"), and finally the Cenozoic ("recent life"). Oversimplifying greatly and focusing only on vertebrates, these three eras might be described as the ages of fish and amphibians; of dinosaurs; and of mammals, respectively.[6]

Each of these eras is further divided into periods. These are the blocks of time most commonly referred to, those that were identified by the earliest geologists in the eighteenth and nineteenth centuries (before the larger blocks could even be imagined). And because these periods were identified piecemeal, over time, their names might seem whimsical or simply obscure. Sometimes they are named after the place where their characteristic rock was first found (Devonian, for example, takes its name from Devon, a county in southern England); or a range of mountains (the now famous Jurassic, after the Jura mountains between France and Switzerland); or what the rocks are composed of (Carboniferous, or coal-bearing); or even from long vanished groups of people (Silurian, from the Silures, a tribe of ancient Britons). These periods, which can span tens of millions of years, are further subdivided into epochs, and ages. From the biggest to the smallest blocks of time, the geological time-scale groupings are eons, eras, periods, epochs, and ages.

Though there is fossil evidence of earlier organisms, the fossil record blossoms when those creatures with hard parts first show up some 540 million years ago, during the very first period of the Paleozoic Era (called the Cambrian, after the name the Romans gave to Wales, Cambria). This is the moment when we see the first clear imprints of the redoubtable sea-dwelling trilobites (early relatives of insects, spiders, and crustaceans).

It was in the Permian Period, which marks the end of the Paleozoic Era, that life came perilously close to vanishing altogether in the greatest of the extinctions to afflict the history of the Earth. From the forms of life that survived that extinction was born the Mesozoic Era, the age of the dinosaurs.

As Ernst Stromer reached the bottom of the escarpment of the Bahariya Depression on the afternoon of January 11, he skirted the flanks of Gebel Ghorabi, one of the many hills and ridges rising from the valley floor, and headed south toward the village of Mandisha. He knew he was walking on rock formed in the Mesozoic Era and, in particular, the youngest period of the Mesozoic, the Cretaceous, which gets its name from the Greek word for chalk, a fine-grained limestone that dominates the marine rocks of this era in many parts of the world. Stromer's British colleague at the Egyptian Geological Survey, John Ball, had already proven that. What Stromer did not know, because no one did at the time, was precisely how old those rocks were in calendar years.

To Stromer, the journey to Bahariya was a natural extension of his successful work in the Fayoum Oasis. The Fayoum was a prodigiously productive fossil area. During the course of the twentieth century it yielded some of the earliest species of apes and other early primates, whales, elephantlike mastodons, strange and immense rhinoceroslike creatures equipped with massive horns, an array of small mammals, giant sea snakes, both long- and short-headed crocodilians, turtles of several types, sharks, rays, lungfish and early catfish, birds, and many more.[7]

But Stromer was, as always, searching for evidence of even earlier ancestors of mammals. It was part of his continuing quest to demonstrate that mammals had evolved first in Africa. The sediment layers at the bottom of the Fayoum Oasis, he knew, were formed in the Eocene Epoch of the Cenozoic Era—relatively recently in geologic terms. Thus, if he wanted to find older mammal fossils than those he had already unearthed in Fayoum, he needed to look in a place where somewhat older rocks were exposed. Stromer knew from the British surveyors that there were older, Cretaceous rocks in Bahariya, so that was the logical next place to look. It was logical because Stromer, like everyone else at that time, thought the Eocene Epoch was perhaps a few million years old and the Cretaceous Period a little bit older.

But he was off—by tens of millions of years. We now know that the Eocene Epoch spanned the Earth's history between 34 and 55 million years ago. But by the time Stromer reached the bottom of the Bahariya Depression, he stood upon rock that modern stratigraphic dating tells us had been deposited nearly 100 million years ago, or between two and three times older than the rocks he had been working on in the Fayoum.

They would yield him little in the way of mammal fossils, for mammals then were small and easily missed. We do not know whether he was disappointed with this discovery. What we do know is that what Stromer found in Bahariya would keep him busy for most of the rest of his professional career.

For Ernst Freiherr Stromer von Reichenbach had walked right into the heart of the age of dinosaurs.

Stromer chose the small village of Mandisha as his base of operations in the Bahariya Oasis, rather than the larger town of Bawiti, for at least three reasons. First, the mamur, or police chief, of the oasis lived here, and Stromer knew he needed to have good relations with the mamur to do his work. Second, the Bahariya Depression is large, and Mandisha was more central than other hamlets to the sites he wanted to explore. Third and perhaps most interesting, given his frequent complaints about the rigors of roughing it, Stromer genuinely preferred the quiet and relative discomfort of camping in less urban areas to the noise and activity of a larger town. He arrived in Mandisha late in the afternoon of January 11 and immediately presented himself to the mamur.

Stromer was surprised to find that the mamur was a blond, blue-eyed gentleman who spoke French. The mamur was surprised to find Stromer in his midst at all, since the authorities in Cairo and at the Geological Survey apparently had never informed him of the paleontologist's impending arrival. His reception, therefore, was cool, but he offered Stromer a place to camp near the home of a doctor recently assigned to the oasis, a thoughtful gesture that gave Stromer an educated and, it would transpire, pleasant neighbor with whom to talk from time to time during his visit. Stromer, exhausted from the journey, ate lightly that night and retired early.

In the morning, the physician delivered hot tea, along with fresh eggs and bread, but Stromer's plan to spend the day resting was hampered by arguments with some of the Egyptians who had accompanied him, by visits from the mamur and his staff, and by steadily worsening weather that included, astonishingly, rain. He had planned to begin exploring the next morning, but before midnight a full-fledged sandstorm blew into Mandisha. "It tears so strongly at the tent," Stromer wrote that night, "that I am constantly fearful of its destruction and get up several times in the night to tie it down." Not that Stromer had much chance of sleeping well that night anyway; his tent and sleeping bag were, as they would be every night during this journey, infested with fleas: "a terrible beginning of my research in the oasis!" he wrote that night.

The next day the storm raged on and Stromer remained in his tent, writing letters to friends and colleagues in Egypt and Germany. The faithful Maslim worked with limited success to clean the sand and dust from their belongings and food provisions, and for most of the day the two cooked, ate, and rested within the confines of the tent while the other members of Stromer's party found accommodation elsewhere in the village. Despite the wind and sand, the physician and an Arabic-speaking clerk from the village visited, and Stromer entertained them with his theories about the ancient history of the oasis and the idea that the Nile had once flowed through the area. He later wrote, with some pleasure, that they were astonished by his theories.

The storm finally abated that night, and on the clear, cool, and finally still morning of the fourteenth of January, after a delay of three days, Stromer at last hiked out of Mandisha with Maslim to begin his exploration of a long ridge called Gebel Hafhuf, in the middle of the Bahariya Depression and southeast of Bawiti. After a long day of climbing, prowling, and documenting the sediments that made up this ridge, Stromer would be rewarded with little more than a fossilized shark vertebra, some fish teeth, the stemlike sections of fossilized wood, and a live scorpion that terrified his companion.[8]

The next day, not unexpectedly perhaps, Maslim complained he was too tired to accompany Stromer, so the dragoman, el Hitu, and one of the camel drivers walked with him instead as he explored yet another major geological feature of the oasis, an elongated oval plateau called Gebel Mandisha, above the village where they were camping. While

Stromer appears to have been happy to chronicle the stratification of the exposed flanks of the ridge, the fossil finds were limited here, too. Over the course of the day he found only a fish vertebra, a piece of what he believed to be shark cartilage, and, toward the end of the day, the fin spines of a primitive shark and what may have been the teeth of a reptile. Tired and disappointed, he returned to his campsite only to find that the mamur and two other civil servants had come to pay him a social call. Politely, he and Maslim served them tea and made small talk until they departed, after which Stromer collapsed in his tent and slept soundly until the following morning.

This day's prospecting, at another remote outcropping, proved so disappointing that by midafternoon Stromer gave up and returned to the area of Gebel Mandisha that had yielded at least meager results the day before. Here he was rewarded by the discovery of several pieces of a crocodile skull. Not far from this spot, he also found a large and well-preserved vertebra he recognized as having belonged to a plesiosaur, a long-necked, paddle-finned marine reptile that reached lengths of up to thirty-five feet. Stromer's patience, it would appear, was beginning to pay off.

The seventeenth of January dawned clear and exceptionally cold. Stromer and his entourage decamped from Mandisha and began journeying north across the oasis floor to the isolated hill called Gebel el Dist.[9] With its almost perfect cone shape and regular horizontal bands of rock strata, el Dist looks for all the world like an immense and elaborate multilayered wedding cake that has been left out in the rain by a distracted bridal party. None of the other hills and ridges in the northern part of the oasis looks even remotely like it. Now, as he neared the end of his visit to Bahariya, Stromer's journals record a heightened level of frustration and anxiety. He had found little so far to justify his long, difficult, and costly journey from Cairo. And while Stromer never seemed to be so happy as when he was scrambling over new exposures of desert outcrops, mapping and describing the sediment layers, it is clear his main purpose for being in Bahariya—to discover new and significant fossils—had thus far been notably unsuccessful. That was about to change.

Just before noon, they made camp in the lee of a south-facing ledge on the west side of Gebel el Dist, and by one-thirty P.M. Stromer was already wandering north along the west slope. Almost immediately, he had some success. In a layer of fine-grained white sandstone, just beneath a brown, iron-rich layer, he found "a few gypsum-weathered bone pieces, among them a crocodile vertebra." Later that afternoon he also found pieces of shark-fin spine. Clearly this was a more promising site than gebels Hafhuf or Mandisha. After spending the rest of the afternoon carefully profiling each of the layers of the exposed face of the hill, which he estimated to have a total height of 492 feet, he returned, heartened, to camp as night came on. The best was yet to come.

On the eighteenth of January, after yet another remarkably rare rainfall, Stromer was up early and picking his way around the south flank of el Dist. He found a variety of bone fragments, but most, to his disgust, were so badly weathered or gypsum-infused that they crumbled when he touched them. But as he wandered among a peculiar group of wind-eroded three- to twenty-foot-high knolls just south of the main hill, he suddenly found "three large bones which I attempt to excavate and photograph. The upper extremity is heavily weathered and incomplete [but] measures 110 cm long and 15 cm thick [about 43 by 6 inches]. The second and better one underneath is probably a femur [thighbone] and is wholly 95 cm long and, in the middle, also 15 cm thick [about 37 by 6 inches].[10] The third is too deep in the ground and will require too much time to recover." Later that same morning, he would find a pelvic vertebra, a flat, riblike bone he identified as an ischium (one of the pelvic bones of a dinosaur), another vertebra with "a convex end"[11] and what he described as "a gigantic claw." All the bones were huge.

With characteristic understatement evidencing none of what must have been his genuine excitement, Stromer wrote in his journal, with what may have been stunned surprise, "Apparently these are the first of Egypt's dinosaurs and I have finally before me the layer that contains land animals."

If his excitement was tempered, it may well have been because he now was faced with a major technical problem: "I don't know how to conserve such gigantic pieces," he wrote ruefully in his journal, "or transport them to Fayoum." After all, he had not come to Bahariya in search of dinosaurs but of much more manageable small mammal fossils. He had neither the tools with which to excavate large bones nor the means by which to carry them back to Cairo and thence to Munich. He looked in vain for smaller pieces of definitive evidence, such as teeth, but complained, "I have little hope of finding smaller remains, especially teeth, because they suffer from weathering, losing their enamel and becoming cracked and shapeless." Nonetheless, he collected one of the vertebrae and a part of one of the large leg bones he found.

That afternoon, Stromer continued around the base of el Dist, moving now to the west side of the steep hill, finding even more large bones, including a possible femur. But this bone, like so much else, was too badly weathered to save. He also found "a gigantic vertebra from a bony fish" and, nearby, "a large black *Ceratodus* [lungfish] tooth." Finally returning to camp, Stromer wrapped his smaller finds in newspaper. Then he cut up a section of his mosquito netting, wrapped the two larger bones he had collected, and coated them in a plasterlike flour and water mixture, following "an American technique taught to me by Markgraf"—one little changed today.

Writing later that night, Stromer reported for the first time, "I am rather satisfied with the results of this day," adding, "and I so enjoy the sound of so many jackals singing in the dark off to the west."

Given his success after only a day at Gebel el Dist, it is surprising that Stromer chose to uproot his team the next morning and trek south again beyond the town of Bawiti to the area near Gebel Hammad. Even on his way out of the el Dist area, he found more fossils—a limb extremity protruding vertically from the ground (which he examined and then covered to retard weathering), a shark-fin spine, and "several armor plates together, the thickest transcending at the surface into brown ironstone, which I take with me since it probably originates from turtles or dinosaurs."[12] But after two days' prospecting around both Gebel Hammad and Gebel Hafhuf, he found little and wrote, "I regret not having remained at G. el Dist." All was not lost, however. Returning to his original base near Gebel Mandisha, Stromer found on January 21 two more intriguing fossil remains, the tooth of a ray, and another vertebra of what he believed to be a plesiosaur.

Two days later, after a day of packing and readying themselves, Stromer and his "people," as he tended to refer to them, were retracing their steps northward toward Fayoum, clearing the summit of the escarpment and leaving Bahariya behind. As they began the trek across the high plateau lands between Bahariya and Fayoum, Stromer was pleased not only by what he had found—a remarkably diverse array of fossil fauna from the Late Cretaceous Period—but also by what he suspected remained to be found. He resolved to engage the services of Richard Markgraf in the following seasons to explore Bahariya further.

The expedition ended where it had begun, in the village of Ghauraq at the southern rim of the Fayoum Oasis. It did not surprise Stromer at all that it took them only six days to return—far less than the outward journey—thus confirming his suspicion that el Hitu had taken longer than was necessary to get to Bahariya. They took the train to Medinet el Fayoum, spent the night, and on January 29 Stromer was aboard an express train to Cairo, spurred along in part by news that Markgraf had been hospitalized there. But by the time he arrived, Markgraf had been discharged and had returned home to Fayoum: two steam engines effectively passing in the night.

In the weeks that followed, Stromer was kept busy by any number of activities, disputes, and geological forays. A rather nasty battle ensued between Stromer and el Hitu as to the matter of compensation—el Hitu wanting more, Stromer offering less, than was originally agreed to because el Hitu had not met the terms of the contract between them—which eventually had to be mediated by the German consul's office. The net effect was that both parties went away unhappy. Stromer delivered a lecture in Cairo on his expeditions. And he made several more short trips of exploration to locations within the Nile Valley, to no great effect. As February drew to a close, and with the help of Markgraf, now recovered, he packed his rock and fossil specimens in eight wooden crates, arranged for their shipment direct to Munich, and made the rounds of his friends to bid them good-bye. He appears to have made a special effort to connect one last time with Herr Rennebaum and his daughter, Elizabeth.

It is clear that despite the hardships of his journeys and his long absence from home, Stromer was not eager to leave Egypt. He made special trips during his last few days to take in some of his favorite gardens in the city and to watch the sun set behind the pyramids at Giza; and he spent one evening simply watching the darkness gather over the Nile as lateen-rigged feluccas ghosted across its surface and the stars appeared.

Finally, on February 18, 1911, Stromer rode the train once again across the flat green delta of the Nile to Alexandria and boarded his Lloyd's steamship for the return journey across the Mediterranean. The crossing would prove rougher than before, and Stromer was frequently seasick, but by February 23 he was home in Munich.

Over the next few years, and especially in the years following the Great War, Ernst Stromer would announce a series of astonishing and unique dinosaur discoveries from Egypt's Bahariya Oasis. They should have made him one of the most famous paleontologists of his era. They did not.

Instead—to the extent anyone remembered him at all—he would be remembered more for what he later lost than what he found.

FIVE

THE ROAD TO BAHARIYA

The jet-lagged members of the Bahariya Dinosaur Project arrived at the Cairo airport a few minutes before midnight on January 11, 2000, eighty-nine years to the day from the date Ernst Stromer arrived in Bahariya. After flying for some twenty-four hours, they found the official welcome at the airport somewhat less than cordial.

"The first thing you see when you get off the plane is a huge sign, in English," recalls Ken Lacovara, forty, an associate professor at the University of Pennsylvania's neighboring institution, Drexel University, and one of the Bahariya Dinosaur Project's field geologists. "The sign says, 'Narcotics Offenses Are Punishable by Hanging.' And you think to yourself, 'Okay, this is definitely a different world.' "

Lacovara became a member of the expedition team almost by accident. He had attended a 1999 talk Josh Smith gave and met with him afterward. Smith, who knew Bahariya had been a coastal environment in the Late Cretaceous, needed an expert in coastal sedimentology, Lacovara's specialty. Lacovara leaped at the offer to participate and would prove a felicitous addition to the team.

For Lacovara, the "different world" of Egypt, a country he would come to love, was about to get even more exciting. Bleary-eyed and weary, the expedition team collected mounds of baggage and tumbled into two small, beat-up white passenger vans, which then tore off into the night.

"I'll never forget that ride," says Lacovara with a look on his face that, with the passage of time, has mellowed from stark terror to mild bemusement. "The two kids who were driving—they looked like they were maybe twelve years old—drag-raced all the way through Cairo, weaving in and out of the traffic at something like eighty miles per hour, leaning on their horns, ignoring lines on the road, and screaming through red lights as if the signals were merely suggestions, which they ignored. At one point the two drivers, going at full speed, were passing things back and forth between them! It was at that point that I decided that wherever else we went on the expedition, I would be the one driving."

The next day, the breakneck speed of the arrival slowed to a crawl. The team had split into two groups. Jason Poole, chief fossil preparator at Philadelphia's Academy of Natural Sciences and the team's fifth principal member, joined project volunteers Jean Caton and Steve Kurth, Penn graduate student Allison Tumarkin, and Patti Kane-Vanni, an attorney and artist with some prior field experience, at an apartment in Giza that had a splendid view of the pyramids. Meanwhile, Josh Smith, Jennifer Smith, Ken Lacovara, Matt Lamanna, and Penn faculty members Bob Giegengack and Peter Dodson moved into the somewhat oddly named Flamenco Hotel in Cairo and embarked on four days of meetings and negotiations with senior officials at the Egyptian Geological Survey and Mining Authority on Salem Saleh Street.

This was a slow, highly ritualized process, as Jen Smith explains: "In Egypt, you don't just storm in, slap down your money, get your permits, and hit the road. You meet. You chat. It's 'How are you?' 'How is your family?' 'What do you think of the weather?' 'Have a seat; have a cup of tea.' It's very civilized. But it takes forever."

"We met with the director of the Geological Museum, Khyrate Soleiman," recalls Ken Lacovara, "and talked and had tea. We were introduced to our three Egyptian partners on the expedition—the geologist Yousry Attia and his students Medhat Said Abdelghani and Yassir

Abdelrazik—and had tea. We met with everyone and had tea. The tea is very hot and very sweet and very strong. It's like magma. I've never been so caffeinated in my life."

"Even before we left Egypt the year before," project leader Josh Smith explains, "Gieg and Jennifer and I had met with officials at the Survey and explained the expedition we had in mind. Then, in the months that followed, we submitted all the applications and security forms and passport pictures the Egyptian authorities required of us. But they hadn't processed any of it until we actually showed up, which I suppose makes sense, since I think they get a lot of requests and a lot of no-shows. So we talked and had tea while all of these things were being processed in various offices. We also needed to acquire maps, military permits for entering the restricted zone in the Western Desert, and permits to work there. It took days."

Meanwhile, between meetings and tea, the project team members went shopping for the tools and supplies they would need during the six-week expedition. The techniques and tools of vertebrate paleontology today haven't evolved much beyond those used by Ernst Stromer and Richard Markgraf almost a century ago. They are decidedly low-tech: brushes of various sizes and stiffness, hammers, chisels, awls, shovels, rope, picks (from as big as pickaxes to as small as dental picks), buckets, water jugs, burlap, and bags of plaster of paris.

"A dinosaur bone," Josh Smith explains, "even a really big one, doesn't come out of the rock in the same condition it went in. A hundred million years is a very long time. The bone is now a fossil, so it's much harder. But it's also much more brittle, and it's usually full of cracks and is very fragile. When we dig it out, we do it very gently and we have to protect it before we can move it, the same way you would cast a broken leg. First we wrap it in paper or aluminum foil, then we cover it with strips of burlap soaked in plaster."

Though they had brought the dental picks with them, they had to locate and purchase almost everything else in Cairo. For a few hours each day, members of the team plunged into the city's chaotic marketplaces.

"Cairo," says Ken Lacovara of their first day of shopping, "is a full-body experience—noises, smells, crowds the likes of which you simply do not experience, I suspect, anywhere else."

Jennifer Smith, on her third trip to Egypt and accustomed to the chaos, says, "Even though Cairo is a major city, there is still that sharp smell of dust from the Sahara in the air. You know the desert isn't very far away. That's how I know I'm back—by that completely characteristic, ancient smell.

"The main marketplace," she continues, "looks positively medieval. There are all these wonderful old buildings with hundreds of tiny shops selling absolutely everything under the sun. Someone will come up to you and say, 'What do you need?' and then, 'Okay, I have it. Follow me.' And you'll follow this guy through these narrow, twisting passages, up stairways—things you would never do here—and eventually he gets you to a place where they have exactly what you need. The people are wonderful and just love it if you try to speak a few words of Arabic. And it's completely safe."

Not everything was easy, however. "It takes a lot of fractured English and Arabic and gesturing," says Jason Poole, "to get across to a shopkeeper that you want his big burlap bags but not the peanuts inside them, and when they figure it out, they just think you're crazy. Which I suppose we were, really."

"Jason was amazing," Josh says. "He would draw sketches and smile and laugh, and eventually it all got done. Of course, when he finally did find us burlap, it was from some old farmer and was full of dried manure and flies, weighed a ton, and was almost impossible to cut. But we had burlap!"

Jason Poole is the sort of fellow who is likely to get attention in a place like Cairo. A big brown bear of a man with shiny dark hair down to his shoulders and a full beard, he is called "Chewie" by the rest of the team, after the similarly hirsute character Chewbacca in George Lucas's *Star Wars* saga—which, for some reason no one can explain, today's generation of paleontologists seems to have committed to memory. Poole, twenty-nine at the time, is also one of the best young bone preparators in America, a master in the delicate art of freeing dinosaurs and other fossils from the rock in which they are locked. Unlike the other four team members, he is an artist, not an academic. Born and raised in Philadelphia, he attended the city's High School for Creative and Performing Arts and later received an associate's degree in commercial art, graduating at the top of his class. Almost as soon as he started working in his field, however, he quit. "I just realized that the commercial art business was a lot more about doing deals than doing art, and I decided I wasn't made that way."

Poole tried a number of jobs that, in one way or another, led him to where he is now. "I taught preschoolers for a while, I taught biology at summer camps, I worked with addicts in the mental health ward of Pennsylvania Hospital. But because I'd been bitten by the dinosaur bug years before, thanks to my father's own interest in fossils, I started doing volunteer bone preparation at the Academy. After a few years I ended up running the education program in the fossil preparation lab, working with kids every day." Today Poole is assistant manager of the museum's dinosaur hall and manager of the museum's fossil preparation lab, one of the few that is open to the public. Jason's lab is run as an educational center in the heart of the Academy's Dinosaur Hall.

"The truth is that people who work with fossils all suffer from arrested development," Poole says, laughing. "We never got beyond the dinosaur stage of childhood. Both my current wife and my ex-wife will be happy to confirm this, I'm sure. I've got a great job; I get to work all day with dinosaur bones and play with kids at the museum.

"And when I'm out in the field with scientists, my mental health training really pays off," he deadpans.

Poole's gentle self-disparagement is charming but utterly misleading. While he doesn't hold degrees in geology or paleontology, he has spent years studying both, often auditing the same courses at the University of Pennsylvania that his academic colleagues have taken.

"Chewie is just a terrifyingly bright guy," says Josh. "The rest of us are academics and, unfortunately, we've been trained to think narrowly. Jason isn't. He's an artist, and an incredibly creative thinker. He's also amazingly quick and funny. If you're trading quips and wisecracks with him, he'll just take you down. And when the tension gets high, he's like a wave that just cuts through and smooths out whatever the problem is. He is completely unflappable. When you combine that with the fact that he's mechanically gifted as well, you couldn't ask for a better field partner."

After days of preparation, a small caravan of white Toyota Land Cruisers, their roofs mounded with lashed-down piles of brown burlap so that they looked for all the world like leatherback turtles on wheels, left Cairo on the afternoon of January 15, 2000, reached the Fayoum-Bahariya road west of Giza, and turned southwest toward the town of Bawiti, 220 miles away in the Bahariya Oasis.

There is no transition between the city and the desert. The suddenness with which the traveler is transported into the unforgiving world of the easternmost extension of the Sahara is like slipping through a crack in the space/time continuum. "After you leave Giza and the pyramids behind," says Jen Smith, "you pass a few scattered housing developments, and even an amusement park called Magic Land. One of the last things you see is a large Coke billboard that says, 'Thirst Is an Ancient Feeling' in English. And then it's the desert. Just like that. Bam!"

In their 1903 survey of the Western Desert, John Ball and Hugh Beadnell described the terrain through which the Bahariya Dinosaur Project team now rode as "a remarkably monotonous undulating gravel-covered desert, the typical 'serir' of the Arabs. . . . Skeletons of camels lie about near the roadside at frequent intervals."[1]

Things have changed only slightly since then. A look at any map of Egypt demonstrates starkly the relationship between water and civilization. Almost the entire population of this nation lives within a few miles of the Nile and its delta. Only about 5 percent of the nation's total land area is settled. With the exception of a handful of oases, most of the rest—including virtually all of the Western Desert, a vast blank spot on the map that occupies fully two thirds of Egypt's land area—is virtually

LEFT: Ernst Freiherr Stromer von Reichenbach as a young boy. BELOW: Stromer, much later in life, a distinguished scientist. *Photos courtesy of Rotraut Baumbaur*

RIGHT: Stromer's wife, Elizabeth Rennebaum Stromer von Reichenbach, holding one of their sons. BELOW: Stromer's three sons, Gerhard, Wolfgang, and Ulman. *Photos courtesy of Rotraut Baumbaur*

Ulman Stromer

Gerhard Stromer

Wolfgang Stromer
Photos courtesy of Rotraut Baumbaur

Verwaltung
der wissenschaftlichen Sammlungen des Staates

Amtlicher Ausweis

für

Herrn Prof.Dr. Ernst Stromer
v o n R e i c h e n b a c h
Abteilungsdirektor a.D.
an der St.-Sammlg.f.Paläontologie

geboren: am 12. XI. 1871 zu
Nürnberg

München, den 19. Februar 1942

Erster Direktor

Eigenhändige Unterschrift des Inhabers

Ernst Freiherr Stromer von Reichenbach

Gültig nur für dasjenige Jahr, dessen Feld (siehe
innseitig) mit dem Dienstsiegel versehen ist.

Stromer's passport. *Courtesy of Rotraut Baumbaur*

The partly destroyed Alte Akademie, home of the Bavarian State Collection of
Paleontology and Historical Geology, in Munich. This building, which housed
the fossils collected by Stromer in Egypt, was bombed during the Allied air campaign in April 1944. *Courtesy of Munich Museum of Paleontology*

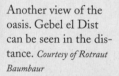

Another view of the
oasis. Gebel el Dist
can be seen in the dis-
tance. *Courtesy of Rotraut
Baumbaur*

Stromer and his cook, Mohammed Maslim, at camp in Bahariya. *Courtesy of Munich Museum of Paleontology*

Stromer and a femur of *Bahariasaurus*. *Courtesy of Munich Museum of Paleontology*

Meter 1 2 3 4 5 6 7 8

A comparative illustration of the skeletons of *Spinosaurus aegyptiacus* and a human, which appeared in one of Stromer's monographs. *Courtesy of Munich Museum of Paleontology*

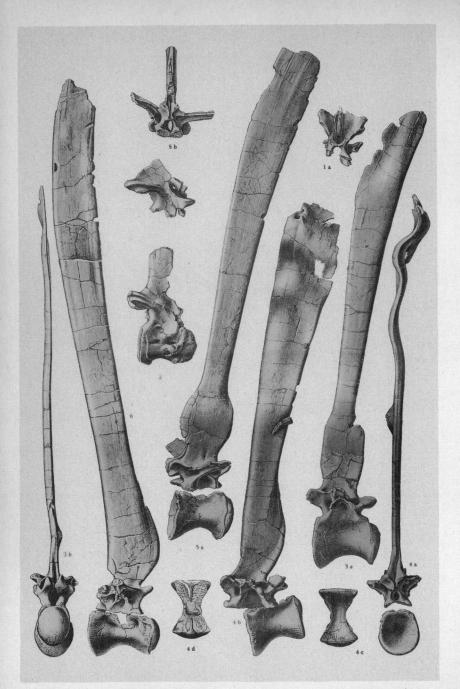

Illustrations of the vertebrate "sail" bones of *Spinosaurus* that appeared in one of Stromer's monographs. *Courtesy of Munich Museum of Paleontology*

uninhabitable. There is only one road into the Western Desert, and the narrow two-lane blacktop to Bahariya is it. It runs nearly arrow-straight southwest from Giza to the Bahariya Oasis. Here the road splits. One branch turns due west to the Siwa Oasis near the Libyan border; the other continues southwest to the Farafra Oasis, where, having reached the edge of the Great Sand Sea, it seems to lose heart. Turning away from the deepening desert, it flees east through the Dakhla and Kharga oases. At Kharga it branches again. One stem turns south toward Sudan, the other turns north and eventually returns to the lush and fertile Nile Valley.

The Western Desert is hyperarid; it is one of the driest regions in the entire Sahara. The climate is unrelievedly inhospitable and the landscape utterly barren in all but a handful of spots. Vast cresting sand-dune belts rise and flow, north to south, like living things relentlessly on the move, eventually reaching deep into the Sudan. The dunes can build as high as 660 feet and stretch unbroken for hundreds of miles. The Great Sand Sea alone is the size of the state of Massachusetts.[2]

But most of the floor of the Western Desert is not sandy. The wind, which is nearly constant, lifts the smallest particles into the air and carries them hundreds, sometimes thousands, of miles. What is left behind is called "desert pavement"—a wind-scoured hardened surface carpeted with gravel, rock, and, not infrequently, fossils—all too heavy for the wind to carry. This is the kind of landscape traversed by the Bahariya road.

"There is a severe lack of green," marvels Jason Poole. "As an easterner, I guess that was what shocked me most. There is no plant life at all. The ground is almost completely flat, and the flatness stretches and shimmers to the horizon. Most of the time there aren't even any mountains in the distance, and when there are, they often disappear behind a sort of fog caused by distant blowing sand. The sky is an astonishing crystalline blue with occasional very clearly defined clouds. Below that blue, though, the world is completely tan, almost gray."

"When you get out and walk on the desert pavement," explains geologist Ken Lacovara, "you notice there is a sort of black patina to it.

The pebbles in this area have a lot of manganese and iron oxide in them, and they blacken with exposure."

"In the middle of all this," says Jennifer Smith, "the road looks completely out of place. And there is also a rail line that runs parallel to the road that carries iron ore from mines near Bahariya to the steel mills at Helwan, near Cairo. Once in a while you'll see ancient ore cars lying on their side where, at some point, they derailed, and no one bothered to right them."

"You still see camel and horse skeletons," adds Poole, "but also rusted-out cars, and you think to yourself, 'Boy, I bet they had a bad day.' "

At roughly the halfway point between Giza and Bahariya, a lone building rises like a mirage along the side of the road. A rest stop. Like everything else, it is the color of sun-bleached khaki. "The place is like the cantina scene in *Star Wars*," says Lacovara. "You run a gauntlet through these packs of feral dogs and step inside, and there will be a handful of mysterious-looking men sipping tea and perhaps a woman behind a veil, and you have no idea how they got there or how long they've been there because there are no other cars outside. It's eerie."

Back on the road, the journey was interrupted once more, by a military checkpoint manned by soldiers with automatic weapons. "To be honest," recalls Matt Lamanna, "the soldiers seemed lonely and happy to see us. Yousry Attia handled the permits and passports, and we were on our way pretty quickly. As we left, I could hear Josh chuckling. Then he said, 'You know those guns they had? They had no ammunition clips in them.' "

Finally, as they continued speeding southwest, the team was treated to what Jason Poole describes as "the longest and most beautiful sunset I've ever seen.

"I guess because of the dust in the air, the colors were just brilliant," Poole recalls. "There was bright purple and vivid orange and neon red. There were colors I have never seen on any palette. And they went on and on for what seemed like forever."

But the only thing moving on the ground was the deepening shadow of the Land Cruisers as they raced southwest across the desert plateau.

It was nighttime when the cars at last descended into the Bahariya Depression and reached Bawiti, the largest settlement in the oasis, but Jennifer Smith had been there before and knew the daytime approach: "The road cuts through a notch in the northeast corner of the depression, and the oasis just opens up in front of you. The escarpment curves away on both sides, and the oasis floor is dotted with isolated round hills and longer linear ridges. The strata—the layers of the earth—are very clear on these hills and ridges. Some of them are capped with ironstone, others with basalt, and still others by limestone. It's the only oasis that has this kind of internal topography. All the others are mostly flat. It's breathtaking."

There are six major oases in the Western Desert of Egypt, transcribing a roughly scythe-shaped arc. From the oasis closest to Cairo and running counterclockwise, they are Fayoum, Siwa, Bahariya, Farafra, Dakhla, and Kharga. They have water, and therefore life, because they lie in depressions far below the otherwise arid and lifeless plateau lands that surround them. They are, effectively, vast deep holes in the ground with sandstones at the bottom from which water bubbles out of perennial springs. Each is an accident of geology. Each is a small miracle.

In Bahariya's case, this water-bearing level, the floor of the depression, is an average of 360 feet below the limestone that forms the cap of the surrounding plateau—the plateau crossed by the narrow two-lane road from Giza.[3] That limestone was created at least 38 million years ago, when a vast sea covered this part of the Earth. Over tens, even hundreds, of thousands of years, trillions of tiny organisms rained their calcite skeletons down to the seafloor where they accumulated, were compressed, and became limestone. Eventually, the sea retreated. Other rock sediments accumulated and later eroded, exposing the limestone to the elements and to the forces of nature.

At some point millions of years ago, this limestone plateau, and all the older rock beneath it, lifted and bowed slightly in the middle along a northeast-southwest axis. Some faulting also occurred, cracking the limestone at the apex of the curved surface. Nature is not kind to such wrinkles in the Earth's crust. It worries them with wind and rain until they lengthen and widen, exposing the rock below. In Bahariya's case, and indeed in the case of each of the oases of the Western Desert, the rocks beneath were erodible. They weathered easily. The cracks not only lengthened and widened but also deepened. The more resistant limestone above was undercut, creating caves. Sections collapsed, creating holes. The holes widened and joined. The scooping action of the wind and, since the Western Desert of Egypt has not always been dry, the weathering work of the rain, with sand as their scouring agent, continued for perhaps 10 million years until they had dug down through the clays that formed the floor of the oasis, exposing, in a few spots, the water-bearing sandstone beneath. More porous than the clay rock that once overlaid them, these sandstones—formed as much as 100 million years ago during the Late Cretaceous of the Mesozoic Era—sprang leaks, and water emerged.

The water in the Bahariya Oasis—and in the Farafra, Dakhla, and Kharga oases as well—comes from a single source: the Nubian aquifer system, a vast expanse of water-saturated rock that lies deep beneath the ground and stretches across most of the Sahara. In Egypt, the aquifer is bounded on the north by an underground interface, effectively a barrier, between the fresh water of the aquifer and saltier water intruding through the rock from the Mediterranean along a curving east-west line running roughly north of Bahariya and south of Cairo. On the east, the aquifer is girded by ancient impermeable "basement" rock that heads north-south for hundreds of miles between the Red Sea and the Nile Valley. There are more varied barriers and connections to the south, into Sudan. But the aquifer has no barriers or boundaries to the west.

For most of the twentieth century, surveyors and hydrogeologists thought that the water trapped in the rock of the Nubian aquifer had percolated north, exquisitely slowly, from equatorial Africa where rainfall is heavy, providing a continuous source of recharge for whatever water was used. In large part because of this theory, the governments of former Egyptian presidents Gamel Abdel Nasser and Anwar Sadat proposed tapping the supply to create agricultural economies in the desert and, not incidentally, relieve population pressures in Cairo, the Nile Valley, and the Delta. But the limited extent and uncertain source of the water under the desert held back these grandiose plans.

It was a good thing. Today, scientists believe the Nubian aquifer is "fossil water."[4] It dates back to periods in the distant past when the region that now is one of the driest on Earth was warm and more humid. Though existing dating technologies permit scientists to say only that this water is at least forty thousand years old, it is almost certainly older—the still puddled prehistoric remains of rain that fell perhaps hundreds of thousands of years ago. Water extracted from this aquifer will not be recharged; when it is gone, it will be gone forever—or until the climate changes again in the region of the Sahara, something that has happened several times before and will probably happen again in some thirty thousand years. Not surprisingly, plans to turn the desert green have been less than successful.

However old it may be, the water of the Nubian aquifer has made Bahariya and its sister oases lushly productive havens in the desert for millennia. Though Bahariya may have been occupied in Paleolithic times, it most certainly supported an agricultural community as early as 3000 B.C. Researchers believe these oasis dwellers were of Libyan origin, and Bahariya itself did not come under the control of the pharaohs for another thousand years.[5] By the time the Romans arrived, in perhaps 30 B.C., Bahariya was already a renowned producer of fine wine (a product that would lose its market when Arabian Muslims conquered Egypt in A.D. 642). It was the Romans who employed their engineering skills to build wells and channels for the springs of the oasis. When the British surveyors John Ball and Hugh Beadnell arrived in the late 1800s, the same springs were still in use, and Bahariya had become a major producer and exporter of dates prized for their quality. The surveyors reported that "though dates are the only fruit exported, olives, apricots, grapes, pomegranates, oranges, lemons, and figs are grown in great numbers, and about 600 feddans[6] are cultivated with rice, wheat and barley crops."[7] More than a century later, many of these same crops are still under cultivation, with methods little changed by the passage of time.

Like the tools of paleontology, field accommodations for paleontologists have changed little since Stromer's time. Dinosaur bones are typically found in remote, dry, badly eroded, not especially hospitable parts of the world—for example, the badlands of Montana or western Canada, the windy wastes of central and southern Argentina, the dusty expanses of western China and Mongolia. The usual place paleontologists call home is a tent. But in the middle of the Western Desert of Egypt, at the bottom of the Bahariya Depression, the paleontologists and geologists of the Bahariya Dinosaur Project were basically in Fat City. They "set up camp" at El Beshmo Lodge, a new, though small, hotel in Bawiti, consisting of a crescent of connected cottages with roofs supported by wooden beams, ceilings made of palm fronds, tiled floors, and walls covered in smooth stucco, like the mud-brick dwellings in the rest of the town. The stucco was painted a warm peach color that glowed in the afternoon, and the whole complex backed up to a lush grove of date palms on the edge of the town. The rooms had beds, bathrooms, and, somewhat irregularly, hot showers. It was, in short, paleontologist heaven.

Except perhaps for the morning wake-up call, which was abrupt, loud, and early. The muezzin's call to worship came over Bawiti's loudspeakers at five-thirty A.M., provoking an equally startling cascade of donkey braying, rooster crowing, and dog barking.

"It was very effective," deadpans Matt Lamanna.

The rude awakening notwithstanding, the view from El Beshmo Lodge that first morning charmed Lamanna: "The grove of date palms behind the lodge was the most amazing green I've ever seen. And it's slammed right up against this completely barren desert. If you step past the date palms, you step from dense greenery to nothing; you have one foot in what looks like a rain forest and the other in what looks like the surface of Mars. The contrast is just awe-inspiring, these ruler-straight lines of irrigated farmland and orchards, and then, with no transition at all, fierce desert."

After breakfast, the team members could hardly contain their enthusiasm to get out into the desert, but Josh Smith called a meeting instead. While he credits Matt Lamanna with the original idea for the expedition to Bahariya, it was Smith's determination that made it happen, and now he had become the expedition leader. It was not a job that sat easily on his shoulders.

"Despite his military training, or maybe because of it," Lamanna says, "Josh is a reluctant leader. He doesn't want to be bossing people around, partly for fear they'll think he's the drill sergeant. But he's really well organized and he sweats the details."

That morning Smith had several details he wanted to get across to his restless team, most of them having to do with safety: Drink plenty of water, use plenty of sunscreen, never go anywhere in the desert without a buddy, never go out of radio range. Then he turned to Jason Poole.

"Because of my work at the Academy, where we also have live animals," Poole explains, "I got to wondering about what sorts of critters we might come across in the desert. Turns out there are three different and dangerous scorpions out there, the yellow fat-tailed scorpion, the black fat-tailed scorpion, and the death stalker scorpion. The last one's name says it all. It's the most venomous scorpion known in the world. There are also foot-long venomous centipedes with stingers at both ends, horned viper snakes, and, my personal favorite, the camel spider, also known as the sun spider or wind scorpion. It is rumored to have a saliva that acts as an anesthetic so that it can eat part of your face off while you're sleeping and you don't even know it. Very nice.

"And the best thing," he adds, "is that you don't even need to leave your bedroom to see a lot of this native wildlife."

Poole told the group they needed to change their normal geological prospecting habits and not stick their hands under every rock that looked interesting. In addition, he made them study photographs so they could recognize these animals, should they encounter one. "That's really important," he explained, "because it's the only way we'll be able to tell the people at the hospital in Bawiti what kind of antivenom to use." Not that the antivenom was risk-free, he commented in passing: "Basically, antivenom is just a different kind of poison. Use enough of it and it counteracts the poison from the stinger. Use too much or too little of it and you die anyway."

Chastened, the group broke up and piled into the Land Cruisers to head out to the desert.

Things did not begin propitiously.

It took the better part of two hours to reach Gebel el Dist, which rises in the northwestern corner of the Bahariya, not far from the edge of the escarpment. But when the Land Cruiser caravan finally ground to a halt at the spot designated by their GPS as the coordinates Josh, Jennifer, and Giegengack had marked the year before, it appeared the landscape had changed completely.

"We wandered around for an hour, at least," Josh Smith recalls, "and part of me was wondering whether the desert could be changed so much by the wind in a year that I wouldn't recognize it. But just viscerally, I knew this was the wrong place. Nothing was right about it. It seemed too high on the slope of el Dist and too far west. But we had the GPS, and it's so incredibly accurate, I was questioning myself. Suddenly I really felt the weight of leading this expedition. I'd brought all these people here. I had a film crew tracking my every move. And I couldn't find the damned site. I remember mumbling to myself, 'Well, I still have my military credentials; I can always reenlist. I don't need to be a paleontologist.' "

Josh should have trusted his instincts. It was the technology that had failed, not his own internal compass. While he wandered off looking for the landscape features he remembered, Jennifer Smith, the GPS guru, and Ken Lacovara fiddled with the equipment and eventually realized what was wrong. At some point in the preceding year, the grid programmed in the GPS had been altered. Reprogramming it as they walked, they arrived at a site Jennifer recognized. And at exactly the same moment, she heard Josh hollering from a short distance away, behind a hill. Following his desert instincts, he had located the same place by sight.

"It was a hell of a relief," he confesses. High-fiving his fiancée, Josh called together the rest of the team. "Now that we'd located the site I'd found the year before, we needed everyone to get their bearings and especially become familiar with what bone in Bahariya looks like—since bone looks different in every formation I've ever seen. In Bahariya, partly because of the high amount of iron in the ground, fossil bone on the surface is oxidized; it looks a lot like charred wood, sometimes with a kind of purplish sheen. But if you look closer, you can see the porosity of the bone, which isn't what wood or rock looks like at all."

Smith's Ph.D. faculty adviser and fellow expedition member Peter Dodson explains further: "The first step in dinosaur hunting is developing a search image. Lots of people can walk across a piece of ground and step right over fossil bone, never even see it. So when we get to a site like Bahariya, the first thing we want to do is look closely at the bone that is characteristic there. We look for the cell structure. If it looks fibrous, you know it's not a rock."

With the search image established, the plan for the rest of the day was a lot more low-tech than its beginning. Armed with little orange marker flags on stiff wire stems, Josh fanned most of the team across the desert floor on the northeast side of el Dist to begin prospecting for bone. Jennifer Smith and Ken Lacovara, the team's geologists, began exploring the flanks of el Dist itself.

Peter Dodson recalls, "It was Medhat Said, one of our Egyptian team members, who found the first fossil, a lungfish tooth plate. But as we spread out, pretty soon bone was coming up all over. It was very promising. We were finding a lot of what I like to call the 'cast of supporting characters'—lots of fish remains, snails, clams, and oyster shells, all sorts of invertebrates."

"The idea was just to identify significant concentrations," Josh says, laughing, "but there was so much bone on the ground that things sort of got out of hand. Towards the end of the afternoon, I had radio reports coming in to me that the team had marked more than a hundred sites. It was crazy."

Says Jason Poole, "I remember turning around at one point near the end of the afternoon, and there was this amazing visual: all these little orange flags fluttering over this dull tannish-gray landscape. And each flag represented fossil bone. There was an incredible amount of bone."

The town of Bawiti to which the team returned that afternoon as the winter light faded probably would have been recognizable to Ernst Stromer. It is, without question, larger today than ninety years ago: The population has more than doubled, to some thirteen thousand people. There are several incongruous multistory buildings here and there, but for the most part, the town remains composed of brown mud-brick single- or double-story dwellings interspersed with domed tombs.

There is a distinctive, and typically Egyptian, pattern to the homes of Bawiti; it is visible to outsiders only from the air. For as long as anyone alive today can remember, entire square blocks of the town have been owned and occupied by large extended families. The plain mud facades that face the dusty streets conceal rather than reveal. And what they conceal is an intimate series of family homes connected by private courtyards. Typically, these courtyards have space for a small kitchen garden and open areas in which children play and women congregate and visit with one another. The tenets of the Islamic culture limit the extent to which women can move freely about the town, but if one chances a glimpse through an open door to the inner courtyards of these domestic compounds, it is clear that there exists a rich interior life.

"If you're lucky," says Jennifer Smith, "you can glimpse the women in these courtyards working with grain or making bread. Often they're wearing brightly colorful clothes. They may have only their eyes showing, but they're wearing hot pink. Only the old women seem to wear black. Behind these walls, which may be centuries old, there is such a continuity of lifestyle.

"You seldom see women in the street, but children are everywhere. The older men would pretend they didn't understand my rudimentary Arabic, but the kids understood me perfectly well. We would talk and laugh together: 'What's your name?' 'How old are you?' 'How many brothers and sisters do you have?' That sort of thing. The kids were terrific."

At dinner that first night, the mood of the group was mixed. Jason Poole remembers, "A lot of us were terrifically excited just to be there at last. Josh was really up because we had made it to Bahariya and had found the sites he'd seen the year before. Peter, who had more experience, was more guarded. But Matt was totally gloomy."

The second thing everyone says about Lamanna is that he wears his heart on his sleeve. A handsome and well-built young man with the wide-open dark brown eyes of a surprised deer and close-cropped hair that is nearly black, Lamanna has the kind of face that telegraphs his emotions instantly, from ebullient enthusiasm to deep disappointment.

The first thing everyone says about him is that he is brilliant. He has an unparalleled knowledge of dinosaurs. Says the Royal Ontario Museum's Hans-Dieter Sues, only half joking: "At the age of twenty-five, Matt Lamanna has already forgotten more about dinosaurs than most of us ever know." He started young. In 1986, when Peter Dodson announced the discovery of a new kind of horned dinosaur, Lamanna clipped the story from his local paper in Waterloo, in New York State's Finger Lakes region, and took it to school. That is where it would have stopped for most young dinosaur enthusiasts, Dodson says. "But Matt sent me this letter filled with the kind of detailed questions I usually expect from college students. He was maybe nine or ten years old. I was amazed and wrote back immediately. I still have that letter."

Josh Smith, who helped recruit Lamanna to do his doctoral work at Penn after the dinosaur prodigy graduated with high honors in geoscience and biology from Hobart College, admits to being in awe of his friend: "Matt can look at some obscure bone that's still coming out of the rock and tell you almost immediately whether it has been discovered before. You can see it happen: His face scrunches up, and you can just see him going through this encyclopedic card file in his head. Then he'll tell you the paper where he read about this bone or one like it, cite the specimen number, tell you the geological formation it was found in and the age of the formation. It's insane. But that's not where it stops. He also has amazing analytic ability. Where Chewie and I might bludgeon a problem into solution, Matt will systematically dissect it until the supports of the problem fall away and it's resolved. It is elegant to watch."

Of the respect in which his colleagues hold him, Lamanna is self-deprecating. "I've been obsessed with dinosaurs since I was four. I've never really thought about being anything else but a paleontologist. I think my experience and my knowledge is pretty narrow, actually; I could use some more interests," he says, laughing.

It may well be that Lamanna's gloom on that first evening was because his mind was racing. "To me, the first day was horrible," he says. "Sure, we found immense amounts of bone, just like Josh said we would, but most of it was incredibly badly preserved, just weathered to pieces and floating on the surface. It was crap."

Largely because they had never worked in North African desert conditions before, it took a while for Josh and Matt to realize that what they were finding on the floor of the Bahariya Depression at the foot of Gebel el Dist was not, for the most part, bone emerging from the rock but bone left behind by the wind.

"In the Western Desert," explains Lamanna, "the wind blows almost all the time and is the major erosional agent. It sweeps away any fine-grained material, including, in time, the solid rock of sandstones and mudstones, but it can't move the bigger material, like chunks of bone. I figured that, since no one had been to Bahariya since Stromer and Markgraf almost a century before, we'd just walk around the hill and find bones in the sand, brush them off, dig around a little, and expose entire articulated skeletons.

"Instead, what we had were these large pieces of bone that were left behind by the wind and, over centuries, had settled to a lower level as the surrounding rock eroded away. They're called ablation lags. And because bones in these lags have been so exposed for so long, they're just beaten to pieces. Combine that with gypsum, a mineral that penetrates every flaw and crack and accelerates the bone's deterioration, and pretty soon you have fossils that explode apart almost the moment you touch them."

Poole recalls, "We didn't talk about it much that first night, but I knew what Matt was thinking because I was thinking the same thing: Most of the bone we'd marked was garbage. I call it 'souvenirosaurus' material. It had no scientific merit at all."

Ernst Stromer had struggled with the same problem in 1911. He complained frequently in his journal about the poor state of preservation of the fossils he was finding. On subsequent expeditions for Stromer, Markgraf would face the same reality: A desert is a destructive environment for bones once they are exposed.

But in 1914, as he eagerly awaited the next shipment from his collector in Egypt, Stromer would discover that poor preservation was the least of his problems.

SIX

FINDS AND LOSSES

The trouble began with a letter from Richard Markgraf. Working for weeks at a time during the winters of 1912 and 1913, Stromer's remarkable bone collector had found, excavated, and recovered a treasure trove of dinosaur bones from the rock and sand of the Bahariya Oasis. On Stromer's instructions, he finally closed the excavations in April 1914 and returned to Cairo to begin the process of shipping the fossils to Munich.

He found Cairo a changed place, however. After years of easy and collegial cooperation, the Anglo-Egyptian authorities now viewed Germans, even trusted scientists like Stromer and collectors like Markgraf, as suspect. In July they refused to permit the shipment to Munich. The Great War, as it soon would be called, was at that moment only a month off.

Markgraf was frantic; he would not be paid until the fossils were delivered successfully to Stromer, and his livelihood, tenuous even in the best of times, was now in jeopardy—months of work, and thus income, held hostage. In retrospect, one cannot help but wonder why Stromer waited so long to shut down the Bahariya dig. The winds of war had been gusting for months; nations had been preparing for it. Perhaps he believed his long-standing friendships in Egypt would survive and even bridge the outbreak of war. Perhaps he believed that science transcended the idiocy of politicians and war makers. If so, he was wrong. Egypt was a strategically vital British protectorate. Given the inevitability of the coming conflict, Stromer's naïveté was stunning. After the outbreak of the war, when it was too late, he wrote to British and Egyptian authorities begging for the release of the fossils, but to no avail. In a sense, it was just as well; there is little likelihood his shipment would have reached Munich after the war had begun.

Then another letter arrived in Munich, this time from Markgraf's wife: Richard Markgraf, never a healthy man in the best of times, had died. His wife was destitute and desperate. Stromer appealed again to his British friends at the Geological Survey of Egypt, and eventually they paid Markgraf's widow a fee and took receipt of the twelve cases of fossil material for safekeeping. Stromer was deeply grateful, writing later that the Survey's action "thus saved [the fossils] from probable destruction."[1] But it would be a full eight years before he would see them.

In the meantime, he wrote monographs on the geology of the Bahariya Oasis and labored to piece together a group of extraordinarily peculiar bones Markgraf had found, excavated, and shipped to Munich in 1912.[2] In 1915, he was at last ready. He published a paper announcing the discovery of *Spinosaurus aegyptiacus*, an enormous predatory dinosaur that probably exceeded *Tyrannosaurus rex* in size. The skeletal parts Stromer assembled, which subsequently were mounted on a wall of the Bavarian State Collection, included the lower jaw, a fragment of the upper jaw, teeth, ribs, and vertebrae from the neck, back, pelvis, and possibly the tail. The body of *Spinosaurus* was, in some respects at least, similar to most other predatory dinosaurs; it walked upright on massive hind legs and had smaller forelimbs, although they were far more robust than those of, for example, *T. rex*, and carried an enlarged claw on the first digit.

Other characteristics, however, marked *Spinosaurus* as peculiar among theropods. Perhaps the strangest thing about Stromer's dinosaur was the vertebrae along its back, which sported bladelike vertical extensions, some more than five feet in length. Nothing exactly like it had ever been seen before. Some scientists have suggested that these spines supported a dorsal hump not unlike that of modern buffalo.[3] Stromer considered this same explanation, but later rejected it as improbable. Today, these spines are popularly believed to have supported a thin skin sail, possibly used by the animal to regulate its temperature.

Strange as this dorsal crest was, however, it was not the dinosaur's most distinguishing feature. That distinction belongs to its teeth, as Hans-Dieter Sues explains: "We distinguish dinosaurs, in part, by their teeth. Teeth vary markedly from species to species and their hard enamel preserves them longer than regular bones. It's really a credit to Stromer's tremendous knowledge of fossil animals that he even identified *Spinosaurus* as a theropod, a meat eater, because *Spinosaurus*'s teeth are completely different from the teeth of the usual carnivorous dinosaur. Most big carnivores have teeth like steak knives; they're long and somewhat flattened side to side and they have serrated edges front and back. But not *Spinosaurus*. Its teeth were long and cone-shaped, almost round in cross section—Stromer called them 'awl-like'—and while they had sharp edges fore and aft, they were not serrated. They weren't for slicing and dicing, they were for puncturing and tearing, like the teeth of today's crocodiles. But crocodiles grab their victims and thrash them side to side. *Spinosaurus* chomped on them and then yanked its head violently up and down, ripping out chunks."

Given these never-before-seen features, Stromer wrote with characteristic understatement, "the establishment of a new genus and species is justified."[4]

Since then, other probable *Spinosaurus* remains have been found in Morocco, Algeria, Tunisia, and Cameroon.[5] Some of this fragmentary material has been assigned to a new species, *Spinosaurus maroccanus*, but scientists disagree about whether there is a clear distinction between the two.[6] More recently, several apparent relatives of *Spinosaurus* have been discovered in Cretaceous rocks from several continents, including *Baryonyx* from England,[7] *Irritator* from Brazil,[8] and *Suchominus* from Niger.[9] Among other things, these latest discoveries demonstrate that the long, low skull of *Spinosaurus* looked more like that of a crocodile than a typical theropod and is one reason some scientists suggest that it may have been a fish eater, though this is disputed by others.[10]

But Stromer's *Spinosaurus aegyptiacus* remains the only spinosaurid yet known to have had extremely elongated dorsal spines. It was, and still is, utterly unique. It should have set the scientific world on fire. It did nothing of the sort. There was a world war on, and it completely escaped the notice of all but the most dedicated researchers.[11]

If Stromer was disappointed, and he could hardly have been otherwise, he did not reveal it in any written document that survives today. There may be any number of reasons why he did not, not least of which was his own pride and professionalism. But he was also otherwise occupied. His medical training had been called upon in the early months of the war, and he served initially as a male nurse. Later he was appointed military geologist at the Geological Survey in Strasbourg, which in those days was in German territory. In a war fought predominantly on land and in trenches, his geological skills would have been invaluable to the war's tactical planners.

Stromer did not return to Munich until November 1, 1919, when he received an appointment to the Bavarian State Collection of Paleontology and Historical Geology. But life was very different in Munich than it had been before the war, and he did not stay long.

Prewar Munich had been a gracious city rich with culture, one of the jewels of the kingdom. It was a wealthy city as well, one that could afford to support the arts and sciences and one in which the level of intellectual discourse was stimulating and vigorous. Indeed, at one point, Stromer was offered the chair of the geology and paleontology department at a university in northern Germany, but he turned it down. He could not imagine leaving Munich.

But after the Great War, Munich was a city at war with itself, a city in which simply surviving from day to day was a struggle. Food shortages had been widespread long before the war ended; now they were catastrophic, and food riots were common. Germany was entering what would become a years-long economic, social, and political tail-spin; what the war had not destroyed, the victorious Allied nations took pains to finish off by diplomatic means. The Treaty of Versailles, signed finally on June 28, 1919, not only altered forever the map of Central Europe, it altered for what must have seemed like forever the lives of the people within it. The Allied demand for reparations payments crippled Germany's already critically weakened economic and financial systems. People were jobless, penniless, and starving. Civil unrest was endemic.

Even as the German old guard struggled to maintain control not just of the country and its economy but of the rigidly stratified society it had created and ruled for so long, radical movements on both the left and the right grew quickly. Workers, emboldened by the apparent success of the Bolshevik Revolution in Russia, formed their own communist party. In fact, even before the Versailles Treaty was signed, the German kingdom of Bavaria—ironically, the most conservative, Catholic, and traditionally nationalistic region of Germany—declared itself the Soviet Republic of Bavaria, with Munich as its capital and a workers' council running the city. The central government had neither the will nor the wherewithal to fight this move. But others did. The end of the war had released a flood of defeated but still staunchly nationalistic officers into a society that no longer revered them and an economy that had no means of supporting them. Searching for someone to blame, they seized upon the communists.

With Bavarian extremists now shooting at one another in the streets of Munich, Ernst Stromer went home to Nuremberg. During the winter semester of 1919 to 1920, he taught at the city's commercial college, and retreated to the family's nearby castle and estate, Grünsberg—where, thanks to its extensive lands and farm, there was at least food to eat and wood to burn in the hearths. Throughout the war and in the hard years that followed, Stromer continued to pursue his research and to publish papers in scientific journals. But the Treaty of Versailles did not bring his exiled Bahariya fossils any closer. Despite the war's end, they remained in storage at the Egyptian Geological Survey.

By October 1920, Stromer was back in Munich with a wife, Elizabeth Rennebaum, and a promotion to the post of chief conservator of the Bavarian State Collection of Paleontology and Historical Geology. Nine months later he was appointed honorary professor in paleontology at the University of Munich, and on July 23, 1921, he was made a full member of the Bavarian Academy of Sciences, a signal honor. Six days later a young Austrian firebrand named Adolf Hitler also got a promotion. He was named chairman of a new political group that called itself the National Socialist German Workers' Party.

It was this latter promotion that would have the greatest impact on Stromer's life.

It was just as Stromer's professional fortunes began to soar that his personal fortunes, like those of so many other German families in the early 1920s, utterly collapsed. As the central government of Germany struggled with the lethal combination of a collapsed economy and punishing reparations payments, it resorted to the simple but disastrous expedient of printing more money. Inflation soared. In the span of four years, the cost of a two-pound loaf of dense German bread rose from a little under three marks to 399 *billion* marks. Citizens staggered home from work carrying suitcases and pushing baby carriages filled with money for food, only to find that either there was none to be had or their "billions" would not pay for even the most basic items. As winter approached, it was literally the case that burning bundles of banknotes created more heat than the amount of coal one could buy with them.[12] The banking system collapsed. What cash Stromer had managed to put away over the years was now worthless. The great Stromer family, while still landed gentry, was effectively impoverished.

For Stromer, it couldn't have come at a worse time. By the summer of 1922, just as the hyperinflation began to peak, his efforts to secure the release of the Bahariya dinosaur fossils from the basement of the Geological Survey of Egypt finally succeeded. Some of the most renowned scientists in the world had come to his aid, including American paleontologist William Diller Matthew and Sir Arthur Smith Woodward, the head of the geology department of the Natural History Museum in London. So, too, had the neutral Swedish embassy in Cairo and the German legation there. As a result of their appeals, the Anglo-Egyptian authorities finally relented. The long incarcerated crates were free to go; the only problem was that Stromer now could not afford to ship them. Deliverance finally came in the person of a former pupil of Stromer's, Bernhard Peyer, an associate professor of zoology and paleontology at the University of Zurich. Having heard of Stromer's plight, he quietly paid some seventy-two British pounds sterling to Cairo officials—a significant sum in those days—to have the crates shipped to his former teacher and mentor.[13] They arrived in the summer of 1922. Stromer was elated.

The elation lasted only until Stromer opened the crates. In Cairo, staff at the museum of the Geological Survey had unpacked and examined the fossils Markgraf had collected. When the time came to forward them to Munich, the museum was slipshod about repacking them. "Hence everything," Stromer later reported, "was rather badly smashed up."[14]

Given his own and Germany's precarious financial condition, and with his colleague and collector Markgraf now dead, Stromer knew he had very little likelihood of ever returning to Egypt for more or better specimens. So he spent the next several years carefully piecing together what was recoverable from the 1922 shipment. Even in its damaged state, the collection would turn out to be, in paleontological terms, a gold mine. In the decade between 1925 and 1935, Stromer (occasionally with the help of colleagues) identified an incredible number of truly remarkable species.

Stromer already had announced in 1914 the discovery of a new crocodyliform (reptiles related to, but not as advanced as, modern crocodilians), *Libycosuchus*, a small, land-based animal that had a skull more like a mammal than a crocodile. Now he added two more: *Stomatosuchus* and *Aegyptosuchus*, and an anomalous pair of specimens they were. For example, *Stomatosuchus*, a huge aquatic beast, had a six-foot skull with an extremely long, flattened snout. Its lower jaw was probably toothless and may have supported a saclike structure not unlike that of a pelican. Stromer identified, but did not name, other types of crocodyliforms in his collection from Bahariya. Two were later named *Stromerosuchus* and *Baharijodon*, but they were based upon very poor fossil material and are currently regarded as invalid genera.

In addition to the crocodyliforms, Stromer identified at least two turtle species, including a side-necked form, *Apertotemporalis*. Incomplete remains indicate that there were a number of plesiosaurs, a long-necked marine reptile, in the Bahariya area as well. And isolated fossils indicate that the primitive, apparently marine-dwelling relative of snakes, *Simoliophis*, was also abundant.[15]

But the most dramatic discoveries from the 1922 shipment were the dinosaurs—with certainty, three more besides the *Spinosaurus aegyptiacus* he had identified in 1915, and perhaps others. In 1931, Stromer announced the identification of a new predatory dinosaur, *Carcharodontosaurus saharicus*,[16] based upon an incomplete skeleton that included several pieces of the skull, teeth, neck, and tail vertebrae, a rib fragment, a hemal arch, fragmentary pelvic bones, and parts of the animal's hind limbs.[17] *Carcharodontosaurus* means "shark-toothed lizard," a name Stromer chose because of the dinosaur's distinctive, nearly straight, grooved, and serrated teeth. A lot more has been learned about this theropod just in the last decade. Remains, principally teeth, have been discovered in Cretaceous rocks in Morocco, Algeria, Tunisia, Libya, Niger, and Sudan. A giant skull, found in Late Cretaceous rocks in Morocco and measuring nearly five and a half feet in length, was announced in 1996[18] and verified that *Carcharodontosaurus* belongs to a larger clade, or group of theropods, known as the *Allosauroidea*, which also includes the well-known Jurassic dinosaurs *Allosaurus* and *Sinraptor*. This new skull, from an animal perhaps 15 percent larger than Stromer's specimen, indicates that *Carcharodontosaurus*, like *Spinosaurus*, rivaled *Tyrannosaurus* in size. *Carcharodontosaurus*'s closest relative appears to be another titanic theropod, *Giganotosaurus*, found in Patagonia. Their similarity probably reflects the fact that Africa and South America had only recently separated, geologically speaking, by the early part of the Late Cretaceous.

In 1932, Stromer described a sauropod from the Bahariya Formation he named *Aegyptosaurus baharijensis*. Plant-eating sauropods, particularly a clade called the titanosaurs to which he assigned this discovery, were abundant in the Cretaceous Period, especially in the Southern Hemisphere. As titanosaurs go, though, *Aegyptosaurus* was only mid-sized, approximately forty-five feet in length—or the size of a small house. Stromer's type specimen consisted of back and tail vertebrae, a shoulder blade, and several limb bones, and was, at the time, one of the most complete skeletons of a titanosaur known. He also found an isolated vertebra from the tail of a different sauropod he tentatively assigned to *Dicraeosaurus*. But that sauropod lived some 60 million years earlier and it may be that the vertebra belongs instead either to a relative, a peculiar tall-spined *Amargasaurus*, known from the Early Cretaceous of Argentina, or to a closely related, recently recognized group of sauropods known as rebbachisaurids—named for a dinosaur first described from rocks in Morocco the same age as those in the Bahariya Formation. Intriguingly, Stromer also noted, among other sauropod material, a single, very large vertebra he could assign to no known genus or species—a vertebra that would figure importantly many years later, in the Bahariya Dinosaur Project.

In 1934, Stromer announced the identification of a third theropod from Bahariya he called *Bahariasaurus ingens*. The type specimen for this discovery included back and pelvic vertebrae, a rib fragment, and three pelvic bones. Stromer assigned other fossils he found in the 1922 shipment to several more individuals of this genus, including a fragment of an ilium, additional pelvic elements, and neck, back, and tail vertebrae. He also tentatively assigned other material, including the femur he collected at Gebel el Dist on January 18, 1911, to *Bahariasaurus*. Taken together, these fossils suggest another large theropod approaching the size of *Tyrannosaurus*. But because of its fragmentary nature, many of today's paleontologists have ignored *Bahariasaurus*, or have suggested it was a close relative of *Carcharodontosaurus*.[19] Then again, in 1996, the paleontologist Paul Sereno identified a new theropod from Morocco, *Deltadromeus*, said to closely resemble *Bahari-*

asaurus.[20] In fact, he assigned the femur Stromer found in 1911, among other of Stromer's fossils, to this new genus—suggesting that it is possible that *Deltadromeus* may also have been present in Bahariya in the Late Cretaceous (though the possibility also exists that *Bahariasaurus* and *Deltadromeus* were the same predatory dinosaur).

Also in 1934, Stromer described several bones in the 1922 shipment as belonging to "*Spinosaurus* B." He believed these bones to have come from two individuals, the larger represented by some vertebrae, the smaller by a few limb elements. Since then, however, the paleontologist Dale Russell identified nearly identical bones from early Late Cretaceous sediments in Morocco and proposed a new genus, *Sigilmassasaurus,* to accommodate both. But Paul Sereno and his colleagues have argued that both belong to *Carcharodontosaurus,* thus rendering Russell's *Sigilmassasaurus* an invalid genus. The controversy has yet to be resolved.

Even with *Spinosaurus* B, however, Stromer had not exhausted the 1922 shipment. In 1934 he also tentatively identified several limb bones to the theropods *Elaphrosaurus bambergi* and *Erectopus sauvergei,* but both are dubious. What is significant, though, is that they document the existence of theropods in Bahariya other than the three—*Spinosaurus, Carcharodontosaurus,* and *Bahariasaurus*—that he identified definitively. In short, it may well be that five—or more—large predatory dinosaurs simultaneously inhabited this remarkable area 99 million years ago, an astonishing number.

The veritable blizzard of new dinosaur discoveries Stromer announced in 1934 should have been front-page news around the world, but was not. There was too much other news on the front pages of the world's newspapers—most notably the tidal wave of misery precipitated worldwide by the collapse of the New York Stock Exchange on Black Friday, October 25, 1929. Germany, still crippled by its reparations payments, was hammered again.

And there was worse yet to come. Germany did not know it in the early 1930s, but the Depression was the least of its problems. Its biggest problem was that its war hero and longtime president, Paul von Hindenburg, now aging and gradually losing his power and place of rever-ence in the hearts of the German people, had been pressured to appoint a new chancellor. The appointment was announced on January 29, 1933. The new chancellor's name was Adolf Hitler.

The National Socialist German Workers' Party had grown dramati-cally during the previous decade, despite its leader's brief imprisonment after an unsuccessful attempt to take over Bavaria by force (the so-called beer hall putsch in 1923). Gradually, through succeeding elections, the Nazis gained a significant minority of the seats in the Reichstag, the German parliament. Hitler's appointment to the post of chancellor was part of a desperate effort by President von Hindenburg's supporters to put together a coalition government that would keep the old guard in power. Their hope was short-lived. Within a month Hitler had seized effective control of the nation, demanding and receiving broad emer-gency powers from von Hindenburg on February 28, 1933. Immedi-ately, he began dismissing, or simply eliminating, his opponents and replacing them with his own associates—names that echo with menace today. Heinrich Himmler became police chief of Munich on March 9, 1933. A month and a half later, Rudolf Hess was named deputy führer of the Nazi Party. Less than a week after that, Hermann Göring, on Hitler's orders, established the Geheime Staatspolizei—the Gestapo.

A year later, President von Hindenburg died. Adolf Hitler immedi-ately declared himself führer of the German state.

One can imagine Ernst Freiherr Stromer von Reichenbach, scion of one of the great patrician families of Nuremberg, longing for the resurrec-tion of a strong and proud German nation amid the chaos and humili-ation that followed the Great War. Initially, at least, he might have found some comfort in the resurgence of national spirit in his belea-guered and virtually bankrupt native country. But Stromer was a man with a sharp, and sharply critical, mind, which saw through the fog of political demaguery just as clearly as it saw through the fog of aca-demia, and he was equally outspoken about both. His critical thinking and plain speaking in his own field of paleontology were legendary. Rudolf Richter, one of the most distinguished paleontologists of the twentieth century and the director of the Senckenberg Museum in Frankfurt am Main, described Stromer admiringly as the most brilliant and knowledgeable of all German-speaking paleontologists.[21] He also characterized him as a man "with a mind that judged without prejudice," who was "inclined to argument almost by principle."[22]

As scrupulously scientific and impersonal as Stromer's journals often are, they nonetheless reveal a man with a passionate commitment to painstaking research, rigorous analysis, and exactitude in scientific description. In an era when new discoveries were the route to recognition and many paleontologists rushed to publish poorly researched and often fanciful papers on their alleged finds, Stromer raged against irresponsible scientific practices among his European colleagues. In a famous speech at a meeting in 1927, he attacked what he believed were twin threats to the science to which he was devoted: *schwindelpalaontologie*, or "fraud paleontology," as he called efforts to bring partial dinosaur skeletons to life in the public mind by embellishing their appearance and habits; and *krümelpalaontologie*, or "junk paleontology," the tendency among some of his contemporaries to claim the discovery of new species from the most inadequate of fossil remains. He was dismissive of both.

But if he was critical of the less responsible of his colleagues, he appears to have been just as critical of himself, often taking his own past work to task for overreaching the evidence. In a congratulatory note published on the occasion of Stromer's eightieth birthday, in 1951, when he had become the most senior of Germany's paleontologists, Richter described him as "a recognized and respected example of what can be done through much sacrifice, individual research and critical consideration, even against oneself."[23]

The most revealing clause in Richter's 1951 appreciation, however, was not that Stromer could be self-critical; for him to be otherwise would have been utterly out of character. The most revealing statement, a phrase that spoke volumes to anyone who knew Stromer well, was "through much sacrifice," for by then Ernst Freiherr Stromer von Reichenbach's commitment to scrupulous personal and professional honesty had cost him dearly—as it did anyone who resisted the Nazi regime. And Stromer did so openly.

In the months following Hitler's August 1934 ascension to the position of führer of the German state, the Nazis moved with extraordinary speed to take control of all of the nation's institutions. The universities were not excepted. At this time, they were instruments of the state, and professors were effectively civil servants. In return for their loyalty, they received life tenure, respectable salaries, and generous pensions. But the Nazis pushed the oath of loyalty to a new and—to Stromer, at least— unacceptable level.

Professors were pressured to join the party. Academics with known connections to the Communist, Socialist, or Social Democratic parties either renounced those connections or lost their positions. But the Nazis' agenda was much more than political. Quietly at first, then more boldly, they purged the nation's institutions of Jews—an act that bled the universities of some of their finest minds.

"You had to prove," explains Hans-Dieter Sues, "that at least back to your great-grandparents you were of pure Aryan blood—which is to say, not Jewish. If you could not, your were harassed and humiliated. In the universities, you were no longer invited to meetings. Your work was ignored. Your funding dried up. As time passed, Jewish professors were dismissed, arrested, or simply disappeared.

"People had two options," Sues continues. "You kept your mouth shut, or you spoke up and expected a call from the Gestapo."

But Ernst Stromer had a third option. He was an aristocrat. In the early years of the Third Reich, the Nazis took pains not to harass the nation's aristocratic families. They understood the power and influence such families still had. As a consequence, individuals from these families enjoyed a certain degree of immunity from the party's worst excesses. But there were limits.

Stromer had become head of the paleontology section of the Bavarian State Collection of Paleontology and Historical Geology in 1930. But there the progress of his career, which had advanced steadily for three decades, slowly ground to a halt. He had refused to join the Nazi Party (though most of his colleagues joined simply out of self-preservation). Worse, he spoke against the Nazi regime and maintained close relationships with his Jewish friends and associates.[24] He was, in short, "politically unreliable."

On July 7, 1937, having passed his sixty-fifth birthday, Ernst Stromer was forced to retire from his post at the university and from his job at the State Collection by the rules of Germany's civil service. In fact, it was amazing that he had lasted that long. His views about the direction the Nazis were taking his country were well known, and only his ancient family lineage protected him from arrest.

Stromer did not retire quietly to the family castle. Instead, he and his family remained in their apartment in Munich, and Stromer remained a fellow of the Bavarian Academy of Sciences. He continued with his research, publishing papers on fossil vertebrates from the region surrounding Munich, as well as critical essays and appreciations of other scientists, including his colleague and friend Ferdinand Broili, the director of the Bavarian State Collection.

By now, Germany was at war with much of the Western world. On March 13, 1938, Hitler annexed Austria. In September of that year, he threatened the annexation of the western region of Czechoslovakia, called the Sudetenland. After agreeing to cease further claims, Germany seized the rest of Czechoslovakia in March 1939. Years earlier Germany had negotiated a nonaggression agreement with Poland; nonetheless, Hitler invaded it on September 1, 1939. Two days later, Britain and France declared war against Germany. World War II had begun.

Closer to home, the Nazis made sure every institution in the nation was headed by party members whose loyalty to the führer was unquestioned. So when Broili reached the mandatory retirement age in 1940, his replacement was Karl Beurlen, a young and well-regarded paleontologist who was also an ardent Nazi. Beurlen would soon demonstrate just how completely, and disastrously, he could be trusted to toe the party line.

Almost from the start, there was a thorn in Beurlen's side, and the thorn's name was Ernst Stromer. As the war deepened, Stromer went repeatedly to Beurlen to demand that the Bavarian State Collection be removed from the museum at the Alte Akademie and placed in a protected location far from the city of Munich. Beurlen rebuffed him dismissively. No less a personage than Luftwaffe chief Hermann Göring had guaranteed that neither Germany nor Munich would be attacked by Allied forces, and Beurlen believed him utterly, the way only a true believer can. As the war progressed, Beurlen's position on this issue became progressively more ridiculous; fine art and science museums throughout Germany were removing their collections to caves and salt mines, but he remained intractable.

In the meantime, Broili (who was still involved with the museum in an emeritus capacity) began quietly to remove small specimens from the museum in his briefcase. Broili had an ally, the paleontologist and noblewoman Princess Theresa zu Oettingen-Spielberg, in whose castle to the north of Munich many of these small, easily removed specimens were stored.[25] But there were obvious limits to what could be transported in a briefcase, and almost all of the collection remained in the museum. With rising insistence, Stromer demanded that the rest of the collection—which, not incidentally, included his own extraordinary finds—be removed. In response, Beurlen threatened Stromer on at least two occasions with his own removal—to a concentration camp.

At some point, it seems clear that Beurlen reported Stromer to Nazi authorities, but Hitler's henchmen took no direct action. They found a more subtle and far more horrible way to punish him.

"In those days," explains Hans-Dieter Sues, "if you had the kind of social standing Stromer did, they left you alone. But your friends might suddenly get in trouble. Or, if you had a son in the military, he might be assigned to a *strafbattalion*, a battalion used for suicidal missions on the front lines, or to some other detachment that was deliberately sent into places where they were certain to be killed."[26]

Ernst and Elizabeth Stromer von Reichenbach had three sons: Ulman, born in 1921, Wolfgang, born in 1922, and Gerhard, born in 1927. Ulman and Wolfgang were sent to the Russian front as soon as they were conscripted.

Ulman was killed on November 10, 1941. Wolfgang survived until November of 1944 and then disappeared in Russia. By this time Stromer and his wife had abandoned Munich and retreated to Grünsberg. But tragedy followed.

It was at Grünsberg that Stromer was to learn of the April 24, 1944, Royal Air Force bombing of Munich, the raid that reduced to ashes his life's work, and everything else in the Bavarian State Collection of Paleontology and Historical Geology that had not been smuggled to safety, just as he had predicted.

Exactly one year later, on April 25, 1945, Stromer's third and youngest son, Gerhard, died within months of being assigned to a battalion fighting rapidly advancing Allied forces in northern Germany. Less than two weeks after Gerhard was killed, Germany surrendered.

Ernst Freiherr Stromer von Reichenbach, never a visibly emotional man, shouldered his multiple losses. His life's work was destroyed. Two of his sons were dead. In Grünsberg, he and his wife awaited word of their missing son, Wolfgang.

It would be a very long wait.

SEVEN

SAND, WIND, AND TIME

The radio message came from Ken Lacovara, who was high on the slope of Gebel el Dist: "Listen, you guys, it looks like there's a sandstorm coming up from the south; better batten things down!"

A few minutes later the storm arrived. It began with an almost imperceptible shift in the normal direction of the wind in the oasis, swinging around to the south from the usual northwest. Then it rose. And as it rose, dust clouds swirled around the paleontologists working on the desert floor. Within only a few minutes the wind was gusting higher and lifting ever larger particles off the desert pavement. Sand devils snaked across the ground, twisting and curling like live things. The colors of the oasis floor, never particularly rich to begin with, gradually turned a nearly uniform light tan, the color of flying bits of quartz. The distant features of the landscape, including the cliffs of the oasis escarpment, softened as if wreathed in smoke, then vanished altogether.

At the foot of Gebel el Dist, the diggers turned their backs to the wind. Wearing ski goggles, with handkerchiefs tied over their mouths, they kept working, picking away at the rock upon which they were sprawled. In what seemed like no time, however, the blowing sand accelerated, rose like a curtain over their heads, and blasted their exposed skin.

"This is nuts," Josh Smith yelled over the wind and the hiss of the sand racing across the desert floor, "I uncover something and in three seconds it's covered again!"

"You'd stand up," Jason Poole explained later, "and you'd be afraid to walk anywhere because you didn't want to step on anything someone had been working on, but you couldn't open your eyes to see. And when you walked, it was like walking through sandpaper. If we'd had shorts on, our legs would have been blasted raw."

Up on the side of el Dist, where Lacovara was trying to work, it was the wind, not the sand, that was the hazard. "It was definitely blowing near hurricane force up there," says Lacovara. "There were several times when I just felt buoyant—which is not a good thing when you're standing on a cliff."

"At one point," recalls Jennifer Smith, who was working with him, "Ken stepped behind me and I nearly tumbled over. I'd been leaning into the wind so hard that when he created a wind shadow, it was as if someone had pushed me from the other side."

Though it came from the right direction, the south, the storm was probably not an early harbinger of the dreaded khamsin, the sometimes fifty-day-long season of frequent sandstorms that typically lasts from March to May. The giveaway was the temperature. The khamsin is a hot wind; this one was cold.

Though the dig had just begun and they'd stuck it out as long as they could, Josh felt he had no alternative but to call off work for the rest of the afternoon. He was as frustrated as Stromer had been when forced to spend two days in his tent for the same reason ninety years earlier. But there was a difference. Stromer was responsible only for himself, the Egyptians being wise enough to seek shelter. Smith was responsible for a whole team. Finally, heads down, buffeted by the wind and blasted by the sand, they abandoned their various sites and fled to the Land Cruisers.

"I've been a field paleontologist for thirty years," says Peter Dodson, "and that was the toughest day I ever spent in the field."

Though the senior member of the expedition team and both Smith's and Lamanna's professor and dissertation adviser, Dodson was not merely an observer in the field. He spent the six-week season on his knees in the sand like everyone else, typically working patiently to finish up sites while his younger colleagues moved on to other locations. His experience helped the team in one other way as well: It dampened their highs and lifted their lows. Says Lamanna, whose highs and lows tended to be the most dramatic, "Peter is the voice of experience; when we were depressed or disgusted about our progress, he'd just gently encourage us to forge ahead. And when we'd get all jazzed about some piece of bone, he'd say, 'Well, okay, interesting, a possible maybe, but let's not get too excited.' "

Peter Dodson would have occasion to be excited himself, but it would take nearly two weeks of digging before he ventured an exultant "Yahoo!"

On the second day of the expedition, the Bahariya Dinosaur Project team returned to the area around the southwestern flanks of Gebel el Dist. Josh Smith explains: "An expedition has four stages: exploration, evaluation, excavation, and—if you decide something is worthwhile—extraction." With more than a hundred orange flags bobbing on the desert floor, it was time to evaluate.

"We all knew, even though we hadn't talked about it much, that most of what had been flagged the day before was crap," Josh admits. "Most of it was float, stuff just left behind by the wind and so badly weathered it was useless. That's one of the things that is so frustrating about Bahariya; there's so much bone everywhere you look, but most of it is so badly preserved that after a while you don't even bend over to look at it anymore."

It was just as well, because the team found that most of the flags had disappeared. "I was thinking to myself," remembers Jason Poole facetiously, " 'Uh-oh, we've offended the park ranger!' " Weeks later, the flags reappeared—decorating the rooflines of a house in the small village of Mandisha, outside of Bawiti, gaily fluttering in the wind. "Actually," says Ken Lacovara, who saw them first, "they looked terrific."

Three of the locations the team had found on their first day at el Dist had shown real promise, and on the second day, Josh Smith divided his colleagues into working groups and they began examining these sites more closely. Ironically, the site where Josh had found the large theropod bone that got the expedition started in the first place was not one of the promising sites. That bone turned out to have been merely float, with nothing else associated with it.

It was Ken Lacovara and Jen Smith who found the first promising site on the first day of this expedition, while setting up the base for the expedition's global positioning system on a low hill not far from the southwest foot of el Dist.

"The top of this hummock," recalls Lacovara, "had weathered to desert pavement, so it was hard and covered with pebbles and what Jen and I realized suddenly were pieces of bone, fish teeth, that sort of thing."

When Matt Lamanna arrived, he didn't even make it to the top of the mound before his eye was drawn to something: "There was this big disclike plate of bone sort of sprawling down the side of the hill, and there was another one higher up, coming out of the rock layer." He immediately began evaluating the site. There appeared to be a possible bone-bearing horizon just beneath the top of the twelve-foot hillock, in a layer of hard mudstone. Matt, Peter, Medhat, Yassir, Patti Kane-Vanni, and Steve Kurth went to work.

While the naming of fossils is a fairly formalized process, the naming of the sites where fossils are found is informal, even whimsical. This first of the promising sites was named by Kane-Vanni after the brand of canned tuna fish the team gathered at the hill to lunch upon that day— an Egyptian brand called "Happy Fish" that appeared to have been manufactured with paleontologists in mind, since it came complete with bones. But the name proved eerily predictive.

When a site looks promising, paleontologists don't just whip out their rock hammers and start pounding away. "Well, we used to," admits Dodson, "back in the old days, in the late eighteen-hundreds, for example. But today we're as interested in where a bone is, and why, as we are in what it is. So we're a lot more careful nowadays.

"The first thing we do is clean up the site, because if it's covered in extraneous stuff, you'll make mistakes and have accidents and maybe damage the fossils. So we get down on our hands and knees and remove loose rocks and anything else that isn't bone. Then we get out our brushes and gently sweep away loose sand and dust. We want to get down to firm rock, and we're always hoping the bone will be coming out of firm rock, or be associated with bones that are, because bone that's still in the rock will be in much better condition than bone that's been exposed to the elements. If we're lucky and we have bone encased in rock, then we get out the hand tools—dental picks and awls—to start chipping away the rock around the specimen."

At Happy Fish, they were lucky. The bone there was still embedded in rock. But precisely what it was would turn out to be a stubborn mystery. Says Dodson, who spent subsequent days excavating this site, "These were clearly large enough to be dinosaur bones, but they were unlike any dinosaur bone I'd ever seen, and pretty quickly Matt and I agreed it had to be something else. The texture on the plate suggested it might have been the shell of some kind of turtle. But Matt, Jason, and Allison argued the material was too thin for that. A case could be made for an amphibian or a marine reptile of some kind. It could have been some kind of fish. I have to say I've rarely been to a site where there were such clearly diagnostic fossils and found myself so singularly clueless about what they were." It would be days before the rock gave up the bones' secrets.

Clearly, Happy Fish was going to be a difficult but potentially productive site. For the next step, Patti Kane-Vanni gridded and began mapping the site. Josh Smith explains: "If we locate a single, isolated bone, we usually just dig it out, jacket it, take it home, and worry about it later. But when we've got multiple fossils, and especially if we have what seem to be pieces of a single skeleton—as we did at Happy Fish and a couple of other sites—we mark off the site with yarn or string in one-meter squares and then systematically evaluate each square of the grid. We'll photograph the site and create a map of each square in three dimensions, identifying the position of every piece we find, from the littlest snail shell to the biggest dinosaur bone. And each fossil fragment we remove gets a number coded to the map.

"This is called collecting taphonomic data—information about what happened to the animal after it died. It's what we need to do to understand not just what we've found but how it got there before we found it. Did the body die right there or did the skeleton get washed there from someplace else? Did it get picked apart by scavengers? Often the clues are sitting right there around the bone. We need to find those clues and then try to figure out what they're telling us. We do it for ourselves and for scientists who will look at this work after us, even a hundred years from now."

While Dodson's group was preparing the ground and beginning to excavate Happy Fish, another group was evaluating a site some fifty yards away on a low ridge where, the day before, Medhat Said had found what appeared to be a lizard vertebra. This was a promising development, since while Stromer had identified some fifty species of vertebrates and invertebrates in Bahariya, he had not found a single lizard. (Dinosaurs, despite the fact that their name means "terrible lizard," are not lizards.) What's more, as Dodson explains, "Lizards were beginning to show some diversity seventy million years ago, but ninety-five or a hundred million years ago—that is, the age of the rock in Bahariya—they would have been very, very early in their evolutionary history. So we really wanted to identify this lizard."

This site, a low plateau perhaps fifty feet long, had an abundance of fossils, but most had weathered severely. "They were horribly preserved," says Matt Lamanna. There were also lots of gastropod, or snail, fossils, as well as fragments of fish bone that had been encrusted with snails before being buried and fossilized.

"For the record," Josh Smith says with frank admiration, "Matt Lamanna disagreed with Peter's lizard notion right away. He identified it right there in the field—and correctly, as it turned out—as the vertebra of *Simoliophis,* a very primitive form of snakelike marine squamate we would later learn was abundant in Bahariya and that Stromer and Markgraf had found as well."

Matt and Peter went back and forth on the subject for days, but by then the site had already become known, rightly or wrongly, as Lizard Ridge. Dodson would eventually ask Medhat and Yassir to scour the site, but first they turned their attention to Happy Fish, which seemed more promising. Lizard Ridge would return to prominence a few days later, when it produced an important fossil discovery.

Matt Lamanna had found the third promising site that first day. He had wandered a bit farther afield, through the hummocky landscape southwest of el Dist, and come upon a large bone emerging from, rather than floating upon, the west slope of one of the mounds in a hard gray siltstone. Since much of the bone so far had been float, and badly preserved float at that, Lamanna was excited by this find. On the second day of the expedition, while Happy Fish was beginning to be excavated, Josh Smith and Jason Poole went to Lamanna's find and examined two bony protrusions emerging from the side of the low mound. They looked to Josh like the peduncles, or joints, between a dinosaur's pelvic bone—an ilium—and the two other bones in the pelvis, the ischium and the pubis. After discussion with Matt, they concluded the bone had belonged to a theropod, a carnivorous dinosaur, and with the help of Yousry Attia, Jean Caton, and Allison Tumarkin, they began excavating the site.

"But that rock was like concrete," Josh Smith recalls painfully. "It was just awful." Both he and Jason Poole bashed the rock around the ilium with pickaxes on and off for days, catching a breath by working at more forgiving sites, including Happy Fish. There, Dodson had found the tooth plate of a lungfish, a descendant of one of the earliest aquatic creatures to develop lungs and thus to survive both on land and in the water. On Lizard Ridge, Josh had found what looked to him like a promising theropod bone. But the highlight of the day was volunteer Steve Kurth's discovery near Happy Fish of what appeared to be the tooth of a *Spinosaurus,* the carnivorous dinosaur Stromer had discovered and that everyone on the team hoped to find to replace the type specimen, or original, that had been destroyed in the 1944 RAF bombing of Munich.

Josh Smith, so high the first day, was less enthusiastic as the second day drew to a close. They had found an interesting array of fossils, but no significant dinosaur remains. "I felt like we were chasing shadows. Time after time this massive, gorgeous accumulation of float turned out to be exactly that, float, and nothing more. It was find a bone, evaluate it, then nothing. Over and over again."

The next day a short distance up the slope of el Dist, on a lumpy ledge overlooking the valley floor, Ken Lacovara and Jen Smith made an exciting discovery.

Lacovara says, "After the rest of the team found those first bones, Jen and I began to trace out that same depositional environment, that same rock layer, around the south side of el Dist, and to follow oxidized bone fragments located there, almost like a trail of bread crumbs. Stromer had called it his *Saurierlagen,* his dinosaur bone layer. Sometime around noon, while Medhat and Steve Kurth were working with us, we reached this ledge about seven or eight feet above the floor of the depression and found much bigger bone pieces."

Lacovara radioed the rest of the crew, and they quickly converged on the site. Josh Smith was delighted, but Lacovara had a bigger surprise for the project leader: three piles of loose, broken rock and sand running in a row north to south; they had nothing to do with the rock structures around them. Lacovara knew what they were immediately: "These are spoils piles left behind by someone else who was digging around the base of Gebel el Dist, and I think we know who that somebody was."

It appeared they had found one of Ernst Stromer's sites, or one excavated a year or two later by his colleague Richard Markgraf. The mounds covered or were adjacent to three backfilled pits.

"What really amazed me," Josh Smith recalls, "was that I realized that Jennifer, Giegengack, and I had sat at that very spot the year before and missed them completely. Now they were plain as day. For the first time, we were in the footsteps of old Uncle Ernst.

"Of course the big question on all of our minds," Josh continues, "was why Stromer or Markgraf—if it was indeed one of them—had gone to the trouble of refilling these pits. The only explanation was that whoever dug them had run out of money or time or both, so they covered them to preserve whatever was in there until they could return. Was it a skeleton? A group of other bones? We immediately began digging to find out."

It was at precisely this point that another one of Bahariya's brutal sandstorms hit. They kept digging anyway, eager to find out whether something had been left behind. Eventually, they did: At the bottom of the three foot-deep pits they found not a skeleton, as they had hoped, but three large, badly preserved dinosaur vertebrae. They had been pedestaled: the process by which the rock around and beneath a fossil is chipped away until all that is left is a slim pillar with the rock-encased bone on top. Even more astonishing, one of them was partly jacketed— that is, someone, presumably Markgraf, had begun covering the bone in newspaper and plaster-soaked burlap, to preserve it for transporting, just as paleontologists do today.

"It was really strange, almost eerie," recalls Jason Poole, who, as the team's bone preparator, often oversaw the jacketing process. "It was as if we were picking up where he had left off." Darkness, made early by the sandstorm, eventually forced them off the mountain. The site they had named Stromer I was left for the next day.

In retrospect, it seems possible that this was perhaps the last excavation ever undertaken by Richard Markgraf for his employer and friend Ernst Stromer in the early spring of 1914. Soon thereafter, World War I intervened, and Markgraf was unable to ship to Stromer even those bones that he had been able to remove; then he had died soon thereafter.

"If it was Markgraf's site, maybe he covered it because he thought he was coming back," says Josh Smith simply. "But never did." And neither did Stromer, who by then was virtually penniless because of the collapse of Germany's currency after World War I.

The crew was back the following morning. If Markgraf had finished the job of jacketing the fossil, the next step would have been to cut away the pedestal beneath, flip the specimen over, and plaster the underside of the jacket so it could be hauled away. Now, though the bones were badly weathered (whether before they had been pedestaled or after was unclear), the team spent the next couple of days finishing the job Markgraf appeared to have begun nine decades earlier.

Fossils—even those partially encased in rock, as were the three sitting on Markgraf's pedestals—are inevitably cracked, brittle, and fragile. They are exposed in the field only enough to determine their significance and state of preservation. For the most part, they are left in the rock in which they are found for protection in transport. Before that can happen, the exposed bone has to be stabilized. These days, Vinac is one of the most commonly used stabilizing agents. Beads of polyvinyl acetate are dissolved in liquid acetone, and the resulting polymer is brushed or dripped onto the bone to create a fast-drying coating that effectively "glues" the fossil so it can be jacketed and moved. The great advantage of Vinac is that it can be removed in the preparation laboratory simply by adding liquid acetone. Without it, or something like it, bones can disintegrate in transit, even in their plaster jackets, becoming extraordinarily difficult three-dimensional jigsaw puzzles—as Ernst Stromer learned only too well in 1922.

Over at Happy Fish, Dodson, Kane-Vanni, Yousry, Medhat, and Yassir continued to work on what now were two large, platelike bones. But, like a frustrated detective, Dodson was no closer to making an identification of the "corpse." As they excavated one of the flat bones, he began formulating an unlikely but pulse-quickening theory: The bone emerging from the hill looked to him like the scutes, or bony scale plates, of a labyrinthodont, a swamp-wallowing amphibian that had been widespread in the Triassic and Early Jurassic periods of the Mesozoic Era. "Could we have one?" he wrote later in large letters in his journal with evident excitement. His excitement certainly would have been justified, for by the time the Late Cretaceous sediments of Bahariya Formation had been formed, most labyrinthodonts had been thought to have been extinct.[1] Finding evidence of one in rock strata this recent would have been an important discovery. In time, however, Dodson would abandon this theory, but others would emerge as regularly as the fossils they extracted from the hill.

Meanwhile, he was busy with another task. One of the principal aims of the Bahariya Dinosaur Project was to strengthen the professional paleontological capacity of the Egyptian Geological Survey and Mining Authority, the project's in-country collaborator, and Dodson spent time teaching his Egyptian colleagues field techniques and, in evening lectures, the finer points of paleontology. Yousry Attia, at fifty, is Egypt's leading vertebrate paleontologist and curator of paleontology at the Geological Museum in Cairo. Having worked closely with paleontological legend Elwyn Simons in the Fayoum Oasis, searching for early mammal fossils, Yousry proved he had developed a sharp eye for small fossils. Almost as valuable, he was exceptionally gifted at helping the team find its way through Egyptian red tape and more than once had to persuade visiting Egyptian Antiquities officials that the project was not plundering archeological sites. Medhat Said Abdelghani, thirty-seven, had a master's degree in archeology but had decided to begin a second career as a paleontologist. Evangelical about the virtues of Islam and of Egypt, Medhat also had a remarkable ability to find fossils. Matt Lamanna has called Medhat "one of the most gifted discoverers of fossils I have ever worked with." Yassir Abdelrazik, despite his youth (twenty-four, the same age as Lamanna), was nonetheless held in high esteem as an intellectual by his older Egyptian colleagues. Typically he rose each morning at two A.M. to study. About to begin his own project studying dinosaur remains in the Mahamid region of Egypt, Yassir had Matt Lamanna as his adviser.

Often it was Yassir who led his fellow Muslims in prayer. At late morning, midday, and late afternoon, the three would withdraw slightly from their dig site, face east, kneel, and perform rituals that seemed to the rest of the crew as timeless as the very desert in which they knelt.

Paleontologists working in close proximity, under often difficult physical conditions, eventually have to find a way to let off steam. On their way back to Bawiti late on this particular winter afternoon, the Bahariya Dinosaur Project found theirs: drive-by fruitings.

Racing across the desert floor in their Land Cruiser, Matt, Josh, Jason, Allison, and Patti launched a vicious ambush on their colleagues Jennifer and Ken, plastering their car with overripe fruit from the oasis. Ken and Jennifer fled across the sand with the other Land Cruiser in hot pursuit, amid much hooting and hollering. It was an ugly, chaotic, and thoroughly delightful exhibition, one that became a regular part of the end of the workday as the weeks progressed, often marked by elaborately planned and messily executed surprise attacks.

The next day, January 20, work returned to normal. Most of the team continued to excavate the Stromer I site, along with another site some three hundred feet away that yielded the fibula (one of two bones in the lower leg) of a large theropod. Then, recalls Josh Smith, "We got a squawk over the radio. It was Ken, saying, 'Hey, you guys better come over here; Jen and I just found lots of really big dinosaur bones just lying on the desert floor.'" Once again, in what had already become a pattern, the geologists had found more new dinosaur bones—though this time somewhat inelegantly, by driving over them with their Toyota.

"I was thinking," says Lacovara, "this is probably the first time a dinosaur has been hit by a car."

Still hoping for a well-preserved, relatively intact skeleton, Josh, Matt, and Jason jumped into their Land Cruiser and raced around the west side of the mountain to the site where Lacovara and Jen Smith waited. When they got there, it seemed to be just what they had been looking for all along: two accumulations of large bones on the surface, seemingly parts of an associated skeleton. The site lay on flat ground a little more than a hundred yards from the low cliffs that formed the base of Gebel el Dist. Lacovara identified the sediments as consistent with some of the other spots where they had found dinosaur fossils. The rest of the team was visibly excited.

"This is what it looks like when you find the real thing," Matt Lamanna enthused.

Recalls Josh Smith, "It was clear to me that this was a big sauropod, maybe Stromer's *Aegyptosaurus*. Our mood went right from the basement to the mountaintop with Ken and Jen's latest find."

But that ebullient mood would not last long. This new discovery created a moment of sharp tension between Josh Smith and Matt Lamanna—both close friends, both excellent paleontologists, both driven to succeed. Smith, unquestionably feeling enormous pressure to unearth a significant skeleton after nearly a week of disappointments, wanted to begin excavating the new site immediately. (Not incidentally, he had learned the day before from film producer Jim Milio that MPH had placed itself in financial jeopardy by financing the expedition, which dramatically increased the pressure to find an important fossil.) Lamanna, something of a compulsive perfectionist, argued that they should finish excavating already existing sites first—particularly the theropod ilium—before they plunged into a new one. Smith did not think much of the ilium site but chose to keep the peace and yielded. Later that afternoon, he, Matt, and Jason Poole, the three strongest members of the paleontology team, took their tensions out on the ironlike rock in which the ilium was encased.

Smith, swinging furiously at the stubborn rock with a pickax, grumbled, only half jokingly, "I want whoever found this fossil killed immediately." The finder, of course, had been Lamanna, and that was all it took to break the tension. The three worked together grudgingly but companionably for the rest of the day. Matt would comment later, "Josh channels his anger really well and takes it out on the rock; it was really a good thing to have him breaking up that rock instead of my jaw."

In the end, however, after all that effort, the ilium proved just one more link in the lengthening chain of disappointments. Though they completed excavating the large bone and jacketed it for transportation back to Cairo, the diggers agreed that it was a poorly preserved specimen and, more disappointingly, that it was isolated—that is, not related to any other bones that they could locate at the site. The ilium site was history.

Says Jason Poole, recalling that day, "None of us was angry at the others, really; we were just feeling beaten down. It was like Bahariya was playing games with us; we'd find a promising accumulation of bone, dig it out expecting to find more underneath, but find nothing. Or stuff too badly preserved to bother with. We'd get excited and then be let down. It was hard not to take it personally."

While Josh, Matt, and Jason were breaking rock, trading brickbats, and getting nowhere, Peter Dodson and his team were still chipping away at the mysterious Happy Fish site. Dodson had run several more theories up the flagpole about the strange, and very large, platelike bones emerging there. One of the flat bones, he suggested idly, might be the humerus, or upper arm bone, of a plesiosaur. The next day another bone emerged from the hillside, one Dodson thought might be the quadrate (a bone in the skull to which the lower jaw is attached) of a theropod, possibly even the long-sought giant carnivore *Spinosaurus*, since they'd also found one of its teeth in the hill. With each new discovery, Dodson, Josh, and Matt would gather and confer, scrutinize and differ, and each time they would decide collectively they still had no idea. But in fact the last theory was very close to the mark. This latter bone was quite clearly a quadrate—but of what?

Meanwhile, across the desert floor at Stromer I, the site where Richard Markgraf had apparently left behind the pedestaled fossils, the rest of the team was finishing the job of plaster-jacketing the bones. The team was working well together now and the expedition had slipped into a familiar, if exhausting, routine. Graduate student Allison Tumarkin wrote in her journal with wry delight: "Okay, now I know I'm really in the field. My clothes are covered in plaster, my hands are permanently Vinaced, my skin is cracking from plaster burn, my knees and shins are covered with bruises and every other word out of my mouth seems to be an expletive. We have so many inside jokes at this point it's like we've invented our own language. How could anyone ask for more?"

That night after dinner, most of the team ended up at a local music concert held in a large round yurtlike structure in Bawiti with a pit fire in the center and straw mats strewn over the sand. Musicians sat cross-legged on the mats—reed instrument players, panpipe players, and drummers playing various forms of percussion instruments. All men. The only women in the crowd were the team's female members, who had come prepared, wearing long clothes and shawls around their heads. The Egyptians in the tent surprised the Americans by dancing sinuously with one another to the mesmerizing beat while everyone else clapped along.

The Bahariya Dinosaur Project is perhaps unusual in that it harbors two drummers: Josh Smith and Ken Lacovara. Lacovara spent a number of years between high school and college playing drums professionally as the house drummer at the Golden Nugget Casino in Atlantic City and in several bands. Having bought a dumbek, a traditional hand-held drum, in Cairo, Lacovara joined in with the musicians, to the amusement of the Egyptians in the tent. "There's sort of a national beat in Egypt," says Lacovara, laughing. "Everyone knows it, whether they're trained drummers or not. It's very fast and very repetitive and there's no room for improvisation. If you pop in a backbeat or start embellishing, they wave their fingers at you, smiling, and say 'La, la, la,' which means No, no, no. For Americans, music—especially jazz—is a form of self-expression. For Egyptians, though, playing in a large group, it's more an expression of solidarity and camaraderie. But it was great fun anyway; I couldn't hold a conversation with these guys, but at least we could speak in music."

The next day, January 23, no one on the team was "speaking music." They were grumbling disgustedly again. In a team evaluation, the latest Ken-and-Jen find, the site that seemed to be an extremely important group of associated bones—parts of a skeleton—was turning into yet another desert mirage, one haunted again by the ghost of Ernst Stromer.

The crew had begun by gridding the site and working on and around the several large exposed bones on the surface, but by noon things had stopped making sense. Josh Smith realized that the sand they were sweeping away from the bone specimens was much younger than the specimens themselves. Initially, he attributed it to erosion. The wind had removed the sand and rock that had originally encased the bones and then, much later, they had been re-covered by newer sand. But while sweeping the southern area of the grid, Matt Lamanna found sand that was much deeper. Switching to a shovel, he soon realized he was emptying a man-made trench nearly five feet deep that had filled up with drifting sand. The trench stretched across the entire back of the site and was bounded by completely unnatural vertical bedrock walls roughly a yard apart. By late afternoon the workers arrived at a depressing conclusion: They were picking over another site that Stromer, or Markgraf working for Stromer, or perhaps someone else, had already excavated. As at Stromer I, they found evidence of plaster jacketing from decades earlier: weathered bits of burlap and crumbling yellow shreds of newspaper printed in German. With little enthusiasm, they named the site Stromer II.

Disappointed again, the diggers turned from the first accumulation of bones at this site to the second, this time with more success. Here they found more than a dozen dinosaur bones, including vertebrae, that Josh Smith tentatively (and, it would later prove, accurately) identified as belonging to a sauropod, possibly the species Stromer himself had identified, *Aegyptosaurus baharijensis*. But all of these bones were shot through with gypsum deposits and crumbling to dust. If this was the site from which Markgraf had excavated Stromer's *Aegyptosaurus*, which did not seem likely at this point, then the collector had taken all the good bones and left behind what Smith's crew was now finding. In Smith's words, "We were finding nothing but ghosts of what once were really good bones. Now most of them were composed of little more than gypsum crystals. I don't know whether that happened before they were excavated or this is simply how fast things weather in the Western Desert, but the bones were just blown apart by gypsum."

In the world of fossil formation, gypsum is an insidious mineral, an agent of quiet but continuous destruction. When water evaporates, it leaves behind mineral crystals. One of the crystals it leaves behind is hydrous calcium sulfate, also known as gypsum. Gypsum is a remarkably useful mineral. Its most well-known and common use is in the manufacture of Sheetrock wallboard, also known to builders as gyp board. In its various forms, gypsum is both the bane and the salvation of paleontologists.

Gypsum exists in the soils of many arid locations (especially in sands) where fossils are often easiest to find. In other, wetter locations, salts like gypsum simply wash away. But in deserts, the salt concentrates. And when gypsum crystals combine with the slightest amount of moisture, a saturated solution is formed that percolates through the sand and infiltrates any porous structure, including buried bone. When gypsum crystallizes out of solution within a dinosaur bone it can, in time, explode the bone. In other circumstances, the gypsum mineral can actually replace the organic material in the bone during the process of fossilization. Either way, it's bad news.

Says Matt Lamanna, "We found bones at Stromer II that had entirely become gypsum. The only way you knew what they were was because their morphology—their shape—had been retained. But the moment you touched them, even just with a brush, they disintegrated."

On the other hand, gypsum makes it possible for paleontologists to jacket fossil bones safely for transportation from site to museum: Plaster of paris is composed largely of gypsum.

Meanwhile, at Happy Fish, the one site that was consistently producing intriguing fossils—though not of dinosaurs—Dodson and Yousry, assisted by Medhat, Yassir, and Kane-Vanni, kept at it. After days of uncertainty and theories, a theme was beginning to emerge. Many of the large, flat bones emerging from the site were discovered to be intricately ornamented, cut with radiating parallel ridges. Dodson had found what he thought were two pieces of the jaw of a fish. Kane-Vanni found two fish scales more than two inches long. Finally, by the beginning of the second week of work on the site, like a camera lens suddenly snapping into focus, the bones of Happy Fish suddenly made sense and were bearing out the wisdom of the site's chosen name: The various large, bony, disclike plates that now littered the floor of the carefully gridded and mapped site were almost certainly once parts of the skull of a fish. But a very large fish. It would be many months before the team was able to definitively identify the fish at Happy Fish as a coelacanth, a lobe-finned fish that first emerged in the Devonian Period of the Paleozoic Era, some 400 million years ago, and became extinct not long after the rocks of Bahariya were laid down. Or at least everyone thought they had become extinct until fishermen off the coast of South Africa brought one up in a net in 1938, a "living fossil." More have been found since and even filmed underwater. As usual, however, Lamanna was ahead of the team. On the day the quadrate first was unearthed, he had dug in to his monographs and found that Stromer had discovered a similar bone and identified it as the quadrate of a coelacanth.

The Happy Fish coelacanth would turn out to be *Mawsonia libyca*. This is a genus better known from Brazil, though, as Stromer proved, not unknown in North Africa. It was a monster, eventually estimated to have been some thirteen feet long. It was, as Dodson later put it, "The sardine that ate Cleveland."

But Matt Lamanna was not impressed. "I did not come to Bahariya to catch fish," he grumbled. In his journal that day, Lamanna would write, "It would seem that the important specimens continue to elude us." To top things off, another harsh sandstorm raced into the Bahariya Depression, and by midafternoon the digs were again brought to a halt and the crew forced to retreat to El Beshmo Lodge.

The next day, January 24, was devoted largely to the less exciting, manual-labor aspects of dinosaur hunting: excavating, pedestaling, plastering, flipping, and finishing the jacketing of the bones that had already been found. As the day wore on, a new problem emerged: One by one, members of the team fell ill—food poisoning or dysentery, no one was ever certain which. By the evening, dinosaur hunting had given way to colleague comforting. In an odd way, the sickness pushed aside the team's dissatisfaction with the dig and the group pulled together again, as healthy members cared for those who were suffering. Whatever their transitory difficulties, this was becoming a very close group of friends and colleagues. Josh called the next day a day off, and the group quietly recuperated from the ailment they called, perhaps inevitably, "Baharrhea."

When they returned to work on January 26, the teams continued collecting and jacketing the fragmentary sauropod bones at Stromer II and excavated the theropod fibula not far from the Stromer I site. At the same time, Dodson's team finished their work at Happy Fish and moved a few dozen yards north to Lizard Ridge, one of the first sites found but set aside in favor of what seemed at the time like more important sites elsewhere. Almost immediately, the new site began paying off. Medhat discovered what looked like another lizard vertebra and another *Spinosaurus* tooth, and Dodson began digging around a complex of bones Josh Smith had found (later determined to be from a very large, bony fish) that had been buried while still being scavenged by hundreds of snails.

In the days to come, the patient work of Medhat, Yousry, and Yassir would yield a number of important discoveries at this site, among them another dinosaur tooth, more *Similiophis* vertebrae,[2] and two standout finds: what looked to be a lizard jaw with well-preserved teeth, and what initially looked like the jaw of a turtle with teeth. Once again, Matt Lamanna, Jason Poole, and Peter Dodson entered into a paleontological debate: Jason held out for a turtle identification, but Matt pointed out what Dodson already knew, that no turtle with teeth had ever been found in Late Cretaceous rock. In fact, no toothed turtles have ever been found in any period. And the fact that no other lizards had been found in Bahariya by either Stromer or Markgraf made Matt certain that the apparent lizard jaw was nothing of the sort. His instincts would eventually prove correct. Months later, Allison Tumarkin discovered that the so-called turtle jaw was from perhaps an even more unusual beast, a strange crocodilian called *Libycosuchus,* a reptile that, unlike most crocodilians, appears to have lived almost entirely on land. The first *Libycosuchus* ever discovered had been identified by none other than Ernst Stromer in 1914.

But these discoveries were still to come. By the afternoon of January 26, when another sandstorm blasted the northern half of the Bahariya Depression, where they had been working for nearly two weeks, the team was happy to abandon their digs. With the lingering effects of illness and the frustrating lack of dramatic discoveries, the energy level was low.

That evening, Jennifer Smith—who had been working elsewhere most of the week with Ken Lacovara—watched her friends drift into El Beshmo Lodge in Bawiti, shook her head in sympathy, and commented, "It's hard to watch these dinosaur guys sometimes. First they find something and it's the greatest thing ever. Then, by the end of the day, it stinks, it's worthless. They go up and down, up and down like they're on a roller coaster. I'm glad I'm a geologist. From one day to the next, the rocks are still just the rocks, telling you stories."

They were stories that stretched back almost to the dawn of time itself.

EIGHT

THE HILL NEAR DEATH

Gebel el Dist is dying. It has perhaps a hundred thousand years to live; not long, geologically speaking, given that it began being created almost 100 million years ago. This fact is so certain you can stand upon it. This fact is not much larger than the average living room. That is the size of the limestone cap that sits atop the 560-foot hill. El Dist is cone-shaped because the sides of the hill are slowly slipping down to the desert floor far below. That the hill exists at all is due to the fact that the limestone cap that sits at its summit—a remnant of the limestone plateau that stretches for miles in every direction beyond the Bahariya Depression— is more erosion-resistant than the rocks beneath it. When the cap goes, el Dist goes. Without the cap, el Dist is washed up—or, to be more accurate, blown down. It's only a matter of time.

"Erosion," Peter Dodson is fond of saying, "is the paleontologist's best friend." It is what releases fossils from the rocks in which they are entombed. But it is very hard on mountains. As the Bahariya Dinosaur Project's paleontologists scrambled around on the dusty, rumpled carpet at el Dist's feet looking for fossils, the most distinctive fossil in the depression loomed directly above them: el Dist—the fossil remnant of millions of years of the history of this particular portion of the planet. Many of the stories el Dist once could tell are gone. Geologist Ken Lacovara estimates that 37 million years' worth of sedimentary layers have long since disappeared—well before the limestone that caps el Dist and the surrounding plateau had even been formed. What remains is really only an outline of the much larger story of Bahariya.

"Here's what the science of geology is like," says Lacovara, who, like all good professors, thinks in easy-to-grasp analogies: "Somebody gives you a hundred loose frames cut from various parts of a full-length feature film you've never seen before, and then says, 'Tell me the plot of this movie.'"

The Egypt movie is very long. Much of it is missing, but what remains is amazing nonetheless. The opening scene, set some 3.7 billion years ago, is stark and terrifying. The Hadean Eon, an interval that lasted nearly a billion years after the Earth's birth, has just ended, leaving behind a barren postapocalyptic landscape. Heavier elements in the still unconsolidated planet, principally iron, have formed a molten core at the center. Closer to the surface, lighter minerals, not yet fully solid themselves, have begun forming a crust. But this crust is far from stable, and during the early stages of the Archean Eon that followed the Hadean, it changed continuously, moving, twisting, turning under convection forces, plunging downward, and slipping upward, but at a pace that was nearly imperceptible, spanning perhaps another billion years. Gradually, sections solidified and began to accrete to form small landmasses. They rose steeply from prehistoric oceans beneath an atmosphere composed of noxious gases and water vapor but no free oxygen, very gradually sliding into one another to form larger masses, breaking apart and disappearing.

Perhaps 2.7 billion years ago, and very possibly earlier, the nucleus of one of these masses emerged in what today is northern Africa.[1] Remarkably, some of these rocks are now visible, in the Uweinat Mountains in the southwest corner of Egypt, in the cataracts of the Nile, and in the Red Sea Hills, exposed again after billions of years. Starting small, this landmass gradually accumulated other bits of crust as it was pushed by tectonic forces over the surface of the globe. From time to time, hot, fresh magma oozed out from beneath this crust, spreading across the landscape. The rocks formed by these events—which continued for another 2 billion years, well into the Proterozoic or third great eon of the Earth's history—are the oldest in Egypt. They form what geologists call the basement complex, the foundation upon which everything else sits. And as with all basements, in most of Egypt these rocks are buried far beneath the surface, under layers and layers of rock and time.

At the same time this infant landmass was growing, it was also moving, carried along on the constantly shifting plates from which the Earth's crust is made. In the Silurian Period of the Paleozoic Era, roughly 420 million years ago—well into the fourth and most recent of the Earth's great eons—Egypt sat not far from the South Pole[2] and was covered with glaciers that began the slow but relentless process of grinding down the earliest mountains. Indeed, geologists have found glacial deposits in southern Egypt from as recently as perhaps 300 million years ago.

Such glaciers would not have lasted long, for the piece of the Earth we now call Egypt, part of the larger piece we call Africa, had begun migrating north toward the equator. And it was not a lone traveler. For one thing, what now is Africa was part of a much larger landmass. For another, there were now five landmasses roaming the globe: two large ones and three smaller ones. The largest have been named Gondwana and Euramerica. The latter consisted essentially of what today are eastern North America and Western Europe, along with Greenland and Scandinavia. The three smaller land areas were what eventually would become southern China, northern China, and Siberia. Africa was part of the largest landmass, Gondwana, along with what today are South America, India, Australia, Madagascar, and Antarctica.

But this, too, was about to change. By the Late Permian Period, the last of the Paleozoic Era, some 245 million years ago, all of these migrating landmasses converged. Now the Earth was composed of only two units: a single enormous continent, called Pangaea, that stretched virtually from pole to pole with Africa at its center, and a single ocean, called Panthalassa, that covered the rest of the globe.

Pangaea was a supercontinent that had come a long way from the barren wastes of the earliest landmasses. Though a mysterious extinction wiped out 95 percent of all marine and land-based animals at the end of the Paleozoic Era, life rebounded by the early Mesozoic Era, just a few million years later. Pangaea teemed and Panthalassa was even more diverse, supporting an amazing array of fish, shellfish, corals, and marine reptiles.

The formation of a single immense continent had profound effects on both the evolution and the distribution of living things. Without the barriers presented by intervening seas and oceans, terrestrial creatures were free to roam and colonize the land. Scientists have found a remarkable degree of homogeneity in terrestrial life-forms across the globe when they examine fossils from this period. But the Earth is a restless place, and neither Pangaea nor that homogeneity was to last long.

That these landmasses were moving at all is something scientists have confirmed only quite recently. Schoolchildren have long noticed that some of the continents, notably South America and Africa, look like pieces of a jigsaw puzzle that once fit together, but it was not until the 1960s that measurement of polarity shifts in the earth's magnetic field, preserved in seafloor lavas, proved that the seafloor was spreading along rifts deep beneath the ocean and that the continents sat upon drifting plates, or slabs, of the Earth's crust and upper mantle, called the lithosphere. In some places, close to where the seafloor was spreading, these plates were growing. In other places, they were diving beneath the continental landmasses, in the process driving up massive mountain ranges by the force of their collision.

At last there was definitive proof of the cause of earthquakes and the tendency of volcanoes to be located in some areas of the globe but not others. At last there was an explanation, amazing but understandable, for why there were fossil shells at the summit of Mount Everest and other mountain peaks: The summits had once been ocean floors. At last too, there was scientific confirmation for why terrestrial forms of life, including early dinosaurs, were similar throughout the globe for some of the Earth's history but then suddenly diverged and diversified. Until the principle of continental drift, or plate tectonics, was established and proved, one widely held explanation for the similarity of terrestrial life-forms on what today are distant continents was that animals had migrated over incredibly long land bridges.[3] Then the sea levels rose, the theory went, the bridges disappeared, migration ended, and species differentiation began.

The notion that entire continents could drift across the surface of the globe seemed preposterous. But it was also a proven fact. What plate tectonics tells us is that this episode of terrestrial species differentiation began, in effect, because it had no choice. As massive chunks of the early supercontinent drifted off across the globe, life began evolving independently in each now isolated location.

But as demonstrable as this theory was, it was also true that no one was certain precisely when these geologic events occurred. To answer that question, scientists—especially paleontologists—study the bones they unearth for clues of differences that distinguish their finds from anything else known on Earth. Dinosaurs, which ruled the land for much of the period during which Pangaea was breaking apart—that is, during the Mesozoic Era—provide tangible clues to the planet's history. Understand when similar dinosaurs began to differentiate, and you begin to have a better idea of when continents may have separated. The reverse is also true: Geologists help paleontologists understand why dinosaur species differ from place to place on the globe by revealing how and when landforms changed.

Thus it was that Josh Smith made quite sure he had two first-rate geologists—Ken Lacovara and Jennifer Smith—studying the layer-cake flanks of Gebel el Dist while his diggers crawled around on their hands and knees down at the mountain's base. The diggers' job was to find the "what" of life in the Late Cretaceous Period of Egypt's prehistory. Lacovara and Smith's job was to explain the "why."

"I was born one house away from a salt marsh in a small town called Pleasantville on the southern New Jersey coast, and grew up in nearby Linwood, a small strip of land with salt marsh on two sides," says Lacovara. "I spent most of my childhood mucking about in the marsh or at the beach. I don't think there were many days when my feet weren't wet. But the thing is, there are not a lot of rocks in that environment, just sand and mud. So when you found any rock bigger than, say, a baseball, it was a remarkable event. My older brother had a rock collection, which may have gotten me started, but maybe not. My mother says even when I was a toddler, she found rocks in my pockets when she did the wash. I picked up whatever I found. When I got a little older, I would go out into our backyard, turn the garden hose on and blast away holes on the grass and dirt, looking for more rocks and checking out the layers when the hole dried out. The yard was sort of pockmarked as a result."

Unlike most of his colleagues on the Bahariya Dinosaur Project, Lacovara's career trajectory took a sharp detour when he fell in love—with a drum set. But after working as a professional drummer for several years, he explains, "I looked around one day and saw all these crusty old musicians rolling home at four in the morning and decided maybe I needed another vocation."

After earning degrees in geography and geology, Lacovara went back to the beach, this time to form his own consulting firm in geology, advising clients on the engineering and geological issues associated with building in or around salt marshes and beach environments. In 1999 he joined the faculty of Drexel University.

His and Jennifer Smith's job on the project was to explore and map the physical landscape of Gebel el Dist and the area around it so the team could better understand the geological history of the area where they prospected for fossils. In the end, it would be Lacovara's knowledge of coastal environments, perhaps even more than Jennifer Smith's extensive knowledge of the Egyptian desert, that would bring this parched landscape to life.

———

The walls of the Bahariya Depression and the slopes of Gebel el Dist offer a graphic, often quite beautiful, though still incomplete picture of what happened to this particular part of North Africa during the Late Cretaceous. But what is visible is only a small part of the sedimentary record of this area of the continent.

Here, as elsewhere on the globe over hundreds of millions of years, the basement layer weathered. The surface was forced upward, fractured, and broken apart; basement rocks with large crystal structures became gravel; rocks with smaller crystal structures became sand. Even finer crystals became silt. Sea levels rose and fell, inundating the land, not once but repeatedly. The lime-rich skeletons and shells of trillions of marine creatures rained down to the seafloor, gently building thick beds of calcium carbonate. Muddy runoff from the land laid down other sediments, finer-grained than the sands and lime layers. From time to time the Earth fissured and molten rock oozed across the surface of the land. In others places, immense quantities of volcanic ash belched into the sky, then settled again on the land and in the sea. The moving continental landmasses collided, buckled, and pushed mountains toward the sky, all with exquisite slowness.

But each time the violence ended, the quieter destruction of erosion persisted and the layering of the Earth continued. During particularly violent intervals, these layers might be fractured and lifted, pushed up and over themselves, even twisted into tortured folds. This process of buildup and breakdown continues even today.

As the Triassic Period ended, the supercontinent Pangaea began to break up, answering to the same tectonic forces that had brought it together in the first place. The landmass called Gondwana remained intact. But the landmass encompassing what today is North America, Europe, and Asia, called Laurasia, began separating from Gondwana along the northern edge of South America and the Moroccan bulge of western Africa. Rifts grew between these two megacontinents, and fresh molten basaltic rock welled up to fill the cavities between the spreading crust segments. Seawater entered the widening rift from the southwest, and the North Atlantic Ocean was born.

Then a rift opened in southern Gondwana, separating Africa/South America from Antarctica/Australia (though a connection may have remained between the southern tip of South America and a corner of Antarctica). The rift grew in a northeasterly direction throughout the Jurassic Period and well into the Cretaceous Period that followed, eventually splitting and separating what is now India from both halves of the divided Gondwana and setting it adrift. In time, this fragment—the Indian subcontinent—would travel northward, collide with the Asian landmass, and, through the force of that collision, drive the Himalayan mountains toward the sky, a process that continues today. Another fragment split off from southern Africa, forming the island of Madagascar.

By the end of the Jurassic Period, another rift began to appear, this time between what now is southern Africa and southern South America. The rift spread northward for the next 50 to 60 million years, gradually peeling South America away from Africa and creating the South Atlantic Ocean. This process continued throughout the middle of the Cretaceous Period, the last period of the Mesozoic Era, until the separation was complete and the South and North Atlantic oceans became one continuous body of water.

While all this tearing and rending was going on, the center of Gondwana—Africa—remained a remarkably peaceful place. There was little in the way of earthquake or volcanic activity, except along the eastern edge, where today's rift valleys now appear. But one event would prove especially significant for Egypt. Beginning in the Triassic Period and continuing throughout the Mesozoic Era, a great body of water called the Tethys Sea grew westward from what is now the Indian Ocean through what is now the Middle East, filling the widening gap between Gondwana and Laurasia. Millions of years later, the Mediterranean Sea is its last remnant. The Tethys was a warm, placid sea, and throughout the immense time frame of its existence, marine vertebrates and invertebrates grew, lived, and died, and thick beds of their skeletal remains blanketed the sea bottom.

The southern shore of the Tethys Sea advanced and retreated repeatedly during this period, often migrating hundreds of miles as sea levels rose and fell worldwide and as the continent itself rose and fell. Meanwhile, the relentless forces of terrestrial erosion continued, wearing down mountains, filling streams and rivers with sediment, and covering the former seafloor with alternating layers of sand and silt. The climate changed throughout the Mesozoic Era as well, alternating between wet and dry, hot and cold. And with each change, the nature and rate of erosion changed. Sometimes water was the principal erosional force, sometimes wind. As each layer built upon the last, the weight of the whole grew, exerting enormous pressure on the strata far below, compressing the sediments into rock: sandstone, mudstone, siltstone, shale, and limestone.

Gebel el Dist, Stromer's isolated conical hill, the stratigraphy of which Jennifer Smith and Ken Lacovara spent several weeks exploring and charting for the Bahariya Dinosaur Project, is effectively a three-dimensional photograph of the history of these ancient layer-building events.

Neither they nor Ernst Stromer were the only explorers to have recognized this. The original British surveyors of the Bahariya Oasis, John Ball and Hugh Beadnell, were the first to measure and define the layered rock strata of el Dist, publishing the results of their analysis in 1903. They were the first to use the term "isolated conical hill" to distinguish el Dist from other ridges and hills in the huge, hollowed-out depression. They identified some twenty-five distinct strata from the sandy beds at el Dist's base to its limestone-capped summit. According to their measurements, the hill was nearly 560 feet high and composed primarily of alternating beds of sandstones, clays, and marls.[4] It was Stromer, however, who proclaimed el Dist the "type section"—that is, the defining example—of the multiple layers of rock that make up the northern escarpment of the oasis, though in truth he could have chosen any section of the area. He called the sequence of rocks the *Baharije-stufe*—the Bahariya Formation.[5]

The word "formation" in this case is a technical term; the complete term is "lithostratigraphic formation." It means a mappable and distinct body of rock (Greek: *litho*) layers (Latin: *strata*) that exist in a certain pattern (Greek: *graphikos*) and will—locally, at least—have the same recognizable relationship with the rocks above and below them, in much the same way as a Linzer torte will always have the same ingredients stacked in the same order. It is this tendency of rock strata in formations to spread laterally and be traceable over distances that made it possible for William Smith to produce that first ever geological map of all of England in 1815. In a sense, Ken Lacovara and Jen Smith were following in his footsteps as well as Stromer's.

Geologists who explored Bahariya after Stromer have further subdivided the Bahariya Formation into different and distinct members, of which the Gebel el Dist member is one. Beneath el Dist—that is, beneath the floor of the oasis—there are another estimated 2,800 feet of additional layers to the basement rock. The Gebel el Dist member is typically described as "made up of fine-grained, well-bedded, ferruginous [iron-rich] clastics [grainy rock] carrying a large number of fossils including vertebrates in the lower levels and an assortment of oysters and other fossils in middle and upper layers."[6]

This is a convoluted way of saying that if you are looking for Stromer's dinosaurs, you need look no farther than the foot of el Dist, in the relatively thin band of rock Stromer called the *Saurierlagen*—the dinosaur layer. And that is where Josh Smith and his intrepid if somewhat disheartened band of paleontologists spent the first two weeks of the expedition.

But Ken Lacovara and Jen Smith had other work to do, and as they prowled around the flanks of the hill, they saw that el Dist held a few secrets. For one thing, Gebel el Dist is an incomplete "photograph" of Bahariya's geological history: Some parts of the picture appear to be missing. And even some of the parts that are there seem fuzzy, out of focus. For example, it has been known for more than a century that almost all of the layers in the Bahariya Formation were formed during what is known as the Cenomanian age, at the very beginning of the Late Cretaceous, which began 99 million years ago. It has long been understood that the various bands of rock that make up the slopes of Gebel el Dist are neatly stacked and firmly compressed layers of sediments that were deposited in shallow estuaries and in saltwater lagoons over a few million years, during which time the Tethys Sea inundated the land and then retreated from it.

It has also long been known that the limestone cap that sits atop these Late Cretaceous sediments was formed during a much more recent period, the Eocene Epoch of the Cenozoic Era, which followed the Mesozoic. And therein lies the mystery: The Eocene did not follow directly after the Late Cretaceous; it did not even begin until 37 million years after the rock in el Dist was formed.

Thus, since erosion happens whenever rock is exposed, either some periods of erosion were more severe than others—severe enough to eliminate nearly 40 million years of sediment that built atop the Late Cretaceous rocks of the Bahariya Formation—or sediments were not deposited during this period at all. Eventually, the Eocene Epoch arrived, bringing with it a warm, shallow sea that remained sufficiently undisturbed long enough for a thick bed of marine skeletons to drift down to the bottom, solidify, and become limestone: the same limestone that today caps el Dist. Presumably, in the 38 million years that stretch back between today and the end of the Eocene, many more layers of sediment were deposited atop this limestone bed, but every single layer that was laid down has since disappeared, for the youngest rock at the top of the Bahariya escarpment is the Eocene limestone. There is nothing younger but the dust and sand that drift and blow across the surface of the desert.

But there was another mystery on the flanks of Gebel el Dist, one that would bedevil Ken Lacovara for days once he began studying the ancient hill. And that was what seemed to him the peculiarly illogical order in which some of the layers were stacked. This was one mystery it would take a while to solve.

If the tools and techniques of paleontology have not changed much since Ernst Stromer's time, most of the tools and techniques of geology haven't, either. It is still basically pick-and-shovel work. Jennifer Smith describes how she and Ken packed for a day on el Dist: "We both had our knapsacks, and besides lunch and plenty of water, we had our rock hammers and a couple of other basic tools, plastic bags to put rock specimens in, indelible marking pens to identify each specimen, and a tape measure. In addition, Ken carried an old army entrenching shovel, the short-handled kind that allows you to adjust the blade to a ninety-degree angle so you can swing it over your head and chop down into the dirt."

Jen Smith also carried one of the team's high-tech additions to this tool kit: the global positioning system, with which they recorded the specific location of the bottom and top of each rock layer they uncovered as they climbed. It consisted of a tall antenna that rose from her pack and connected them to the base station erected at the Happy Fish site, a battery pack strapped around her waist, and a handheld computer for data.

On their first day at Gebel el Dist, Lacovara and Smith circled the hill in their Land Cruiser, taking the measure of the mountain. "We were trying to find the best and safest route to the summit and the area where the rocks were best exposed," Lacovara explains. "That turned out to be the west side, and that's where we spent the first three weeks." Having chosen a route, the two climbed to the summit.

"At five hundred and sixty feet," says Lacovara, "el Dist is about the height of a fifty-six-story building, and it's a tough climb. You're climbing on steep slopes that are often made of loose, sandy scree that gives way as you climb. It's like trying to walk up a down escalator; your quads get quite a workout. There are also some vertical cliff faces, including the summit cap, that I wasn't sure we'd be able to climb without ropes, but Jen wanted to climb the mountain, so that's what we did. And the view of the oasis from the summit was just breathtaking."

The next day, their "tourist" stage over, they went to work. Says Lacovara, "The first time you go into a new geologic area, everything is kind of a blur for a few days. It's like going to a party, meeting thirty people, and trying to remember their names and what they do. You can't do it. You want to get to the stage where it's like a class you've been taking for three months and you know everyone's name and where they sit and their various little habits. That's where we needed to get. So we did what all geologists do: We started at the bottom and worked our way to the top."

Lacovara and Smith's job was to examine and map each layer of el Dist, from the lowest to the highest, in as much detail as the rocks themselves would permit. For most of el Dist, this was not a simple task; with the exception of a few bands of hard rock with vertical cliff faces, most of the slope was covered with material that had eroded and slipped down from above. This is where Lacovara's trenching shovel came in. "El Dist," he says, "is more of a shovel environment than a rock-hammer one."

"Ken was just a digging maniac," Jen Smith says, laughing. "Starting right from the bottom the first day, he would straddle the slope in front of him and start swinging furiously away with his shovel with the blade at a ninety-degree angle until he had exposed a nice, clean vertical or sloped section a yard or so high. He looked like a giant prairie dog."

While Lacovara was burrowing away, Smith walked laterally around the area where they were, logging location and elevation data into the GPS. "There aren't any really detailed maps of the topography of this area; with a few more years of collecting data, we should be able to produce one."

When Lacovara finished digging a section, the two geologists would get down on their hands and knees to scrutinize the clean face. They would determine what kind of sediment they were looking at, what its structure and mineral content was, discuss how it had come to be deposited in this particular spot, take a sample, and try to figure out where one layer ended and another began.

"That's not as clear as you might think," says Smith. "Sometimes the distinction jumps right out at you, maybe a hard brown mudstone layer sitting cleanly on top of a lighter sandstone. But you also could be looking at sand with a bit of mud in it and find above it mud with a bit of sand in it, and you have to try to decide if that's a new layer or just a continuation of the other one. It can be tricky."

At some elevations, as the days went on and the geologists worked higher, this task was easier. "There are places," recalls Smith, "where the sandstone is ferruginous; it's basically ironstone, and it's hard as iron, and all you need is the pointed end of the hammer to clean a surface. In other places there are thick blocks of mudstone where you can scrape the weathered material off and get a good sample of clean rock. So Ken didn't always have to bust into the rock itself, just clean it off."

Before the invention of the GPS, geologists had to measure the length of a sloped section with a tape measure, determine the angle of the slope, then calculate the vertical thickness of the deposit. "Today," Jen Smith explains, "the GPS satellite sends a signal to our base receiver down at Happy Fish and gives us the precise elevation and location where I am standing. To measure the thickness of a sloped layer of rock now, I just stand at the bottom of it and take a reading, then stand at the top and take another reading. The difference is the vertical thickness, accurate to within twenty centimeters [about eight inches]. Later, we can create equally accurate maps in three dimensions simply by plotting out all the readings we take in the field."

There is nothing inherently interesting about a rock. Or a layer of rock. What fascinates most sedimentary geologists—as distinct from, say, children, who will pick up a rock simply because it is there—is that the rocks have stories to tell about what a particular part of the world looked like at the moment the rock was formed. That is what geologists live for: to unravel those stories. In the case of the sedimentary rocks that form the Bahariya Formation, the way Smith and Lacovara hoped to find out about the landscape once roamed by the animals their colleagues were unearthing at the bottom of the hill was by studying something that might well seem crashingly dull but turns out not to be: the size of the grains in the rock.

It is quite simple, really. Sedimentary rock is formed when particles of rock are transported from one place (their place of origination) to another (their place of deposition). The only difference between the clay an artist uses for sculpture and the rocks a builder uses to craft a wall is particle, or grain, size. There is a formal scale by which such things are sorted, called the Wentworth Scale. Clay is defined as any material with a particle size smaller than thirty-nine one thousandths of a millimeter (0.0039 mm). A boulder, which one might think was an informal designation for a big rock, is any particle of rock larger than 256 millimeters (about ten inches) in diameter.

Grains of rock are transported in three ways, depending on how large they are. Fine-grained materials—mud, silt—are transported in suspension; that is, suspended as separate particles, typically in a fluid, more often than not water but sometimes air. Medium-grained materials, like sand, are also carried along by this fluid but tend to bounce along the bottom instead of being suspended, a process called saltation. Boulders simply roll, a process called traction. The faster the fluid is moving, the larger the particles it can carry, and the farther it can carry them. But when the fluid slows, the particles slow as well. The bigger and heavier ones stop first. A flash flood can carry boulders the size of automobiles (and automobiles as well, for that matter) substantial distances. A fast-flowing mountain stream can carry significant amounts of rock and gravel much farther, until it reaches the valley floor below, slows, and begins to release the debris it carries. The finest particles keep moving until the fluid comes to a halt. To use an extreme example, anything carried by water or wind to a lake, which has no current at all, simply settles to the bottom, including the finest silts—which explains why most swimmers don't like putting their feet down on a slimy, fine-grained lake bottom but have no reservations at all at a sandy beach.

When Lacovara and Smith break a piece of dark brown mudstone from an outcrop on the face of Gebel el Dist, for example, they know a few things immediately. First, since the particle size of the grains in a mudstone is very small, the depositional environment in which the mudstone was formed had to have been still water, otherwise the tiny particles would have kept right on going. Second, depending upon how thick the layer of mudstone is, they can hazard a guess about how long this particular spot was placid. Third, they can assume that the particles

LEFT: Jen Smith with her global positioning system. *Courtesy of Vladimir Perlovich*
BELOW: Bahariya Oasis. *Courtesy of Ken Lacovara*

Members of the Bahariya Dinosaur Project team at the Paralititan quarry. In the center of the photo, already jacketed in plaster for the trip back to Philadelphia, is the giant humerus. *Courtesy of Allison Tumarkin*

LEFT: Matt Lamanna at the Paralititan quarry. At his left is the left humerus. BELOW: Matt Lamanna wipes sand away from bones. *Photos courtesy of Vladimir Perlovich*

ABOVE: Josh Smith and Allison Tumarkin working the site. *Courtesy of MPH Entertainment* RIGHT: Jason Poole, aka "Chewie." *Courtesy of Vladimir Perlovich*

ABOVE: Matt Lamanna at South Sauropod site. At his left is the humerus bone. *Courtesy of Ken Lacovara*

RIGHT: Matt Lamanna and Chewie jacketing the bone with plaster and burlap. *Courtesy of MPH Entertainment*

Peter Dodson, with the jacketed humerus. *Courtesy of Bob Walters*

TOP LEFT: Jen Smith and Ken Lacovara at the base of Gebel el Dist. *Courtesy of MPH Entertainment* TOP RIGHT: Josh and Jen Smith in the Everglades. *Courtesy of Ken Lacovara* RIGHT: Ken Lacovara in the Everglades. *Courtesy of Ken Lacovara*

The Everglades represents the closest present-day approximation to the ecosystem of the Bahariya Oasis 95 million years ago. *Courtesy of Ken Lacovara*

TOP: A rendering of what *Paralititan stromeri* would have looked like.
BOTTOM: A lateral view of *Paralititan*'s skeleton.
Illustrations courtesy of Bob Walters and Bruce Mohn

TOP: A rendering of what *Spinosaurus* would have looked like.
BOTTOM: A lateral view of *Spinosaurus*'s skeleton.
Illustrations courtesy of Bob Walters and Bruce Mohn

© 1997 Robert F. Walters

Mohn 2000

TOP: A rendering of what *Carcharodontosaurus* would have looked like.
BOTTOM: A lateral view of *Carcharodontosaurus*'s skeleton.
Illustrations courtesy of Bob Walters and Bruce Mohn

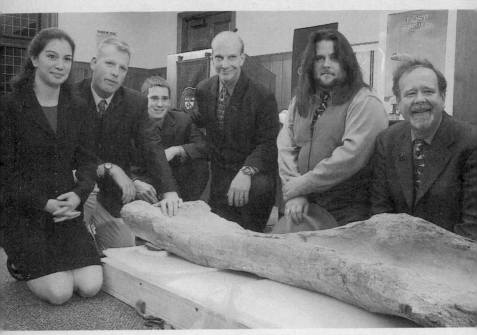

Left to right: Jen Smith, Josh Smith, Matt Lamanna, Ken Lacovara, Jason Poole, and Peter Dodson. In front is the humerus bone of *Paralititan stromeri*.

Courtesy of MPH Entertainment

from which the rock is formed probably originated somewhere far away (in this case, far to the south), since the tiny particles of mudstone can be carried so easily over long distances.

Sandstone is a different story. It is composed of much coarser grains of rock, which can be deposited even when water is moving. The coarser the grains of sand left behind, the faster the water was moving; the finer the grains, the slower the water was moving. By looking hard not just at the grain size of a sandstone, but the way the bed has been laid down, you can tell a lot about the character and history of the place where it was formed millions of years ago.

"The thing to remember," says Lacovara, "is that every layer of sedimentary rock was a modern depositional environment at the time it was laid down, and each depositional environment has a signature set of processes that operate there. Waves deposit sediment differently than rivers do. Wind deposits sediment differently than waves do. You look for their unique signatures. And when you decipher the writing, there's a story that it tells.

"But if we're going to understand the whole story, we need to understand the succession of each layer. We know that only some kinds of depositional environments can come before other kinds. We know only some kinds of environments can follow other kinds, unless there has been some kind of a break—an unconformity—in the geologic record. When we find strata lying above or below each other that violate our understanding of the likely succession, we know we have a mystery to solve."

One of the ways of dividing sedimentologists, or scientists who study sedimentary rocks, is by the depositional environment in which they specialize. There are fluvial sedimentologists, who study ancient river systems; marine sedimentologists, who study ocean bottoms; and coastal sedimentologists, who study the rock formed where land and water meet. Jennifer Smith, for example, is primarily a fluvial sedimentologist; she specializes in riverine environments. Ken Lacovara, as perhaps was inevitable given his childhood fascinations, is primarily a coastal sedimentologist.

And one day, perhaps a third of the way up the side of Gebel el Dist, Lacovara saw a world within the rock that he recognized immediately: beach sands.

"We scraped away a lot of loose stuff beneath a ledge, and right before me were several beautifully formed, cross-bedded layers of sand, the kind of beds—some sloping one way, others sloping another way—that you only see where a beach has been for a very long time. And I realized I was standing on a beach formed almost one hundred million years ago at the edge of the Tethys Sea.

"It may seem odd, but that was a very emotional moment. Here I was standing in the middle of the Egyptian desert, but in my mind's eye, I could see the water of the Tethys Sea stretching away from me toward the north. My feet would have been in the shallows along the beach. I could even tell that the water would have been lapping at my feet very gently. There would have been no crashing surf. The grain size of the sandstone was too fine. If there had been surf, or any significant current at all, the fine-grained sand would not have ended up here beneath my feet. This would have been a quiet, peaceful spot by the sea."

Nearby, Lacovara and Smith found more evidence that they were at the edge of the ancient sea. Beneath one eroded ledge, they looked up and could see plainly the rippled bedding of a sandbar. The sand layers had a greenish tint produced by a mineral called glauconite, almost always the sign of a marine or coastal environment. And just above the rippled beach sand was a shell bed thick with the remains of *Exogyra*, a type of oyster. The ripples themselves, preserved in rock, were a small miracle of geology.

"As you probably know from your own experiences," Lacovara explains with awe, "ripples in a sandy beach are created in a matter of a few seconds, as a small wave retreats or the tide goes out. When Jen and I looked at that rock, we were looking literally at a ten-second snapshot of part of the history of the world, a moment almost a hundred million years ago—one frame of the movie—preserved for all time."

"It's moments like these," says Jen Smith, "that make it so rewarding to be a geologist. You have this sense of connectedness through geologic time that most people never get to experience. It's pretty awesome. I have picked up a rock that was three hundred million years old that had raindrop impressions in it; I held in my hand one moment in time, frozen in the rock. It is just awe-inspiring."

Apart from the excitement, Lacovara was also relieved. "Just as it was a big relief for the paleontologists when they found the first dinosaur bones, it was a big relief for me when I found the first beach deposits. I knew where I was, what kind of system I was in, I knew I could explain this world those dinosaurs lived in."

But in the days to come, Lacovara would discover that the explanations were harder to divine than he first imagined, the mysteries in the rock more difficult to unravel. Until, at least, he had the breakthrough vision that explained it all.

NINE

SOLVING STROMER'S
RIDDLE

"Chewie, dammit, get out of my light," said Josh Smith, playfully shoving Jason Poole away.

"No way! You get out of *my* light," Poole shot back.

On the afternoon of February 3, 2000, Poole and Smith were sprawled a few feet apart on the ground in a particularly barren part of the Bahariya Depression's desert floor. They had been working in this spot, a new and promising site a couple of miles southeast of Gebel el Dist, for most of the day, as they had been for several days. After clearing away loose sand and gravel with stiff bristle brushes and doing a preliminary reconnaissance of the site, they and some of their colleagues had chipped away rock with hammers and awls, sometimes switching to dental picks, and had exposed a number of large dinosaur bones belonging—they were almost certain—to a sauropod, a plant-eating dinosaur from the Late Cretaceous. On this particular afternoon, with the winter light beginning to fade, Smith and Poole were working their way carefully along what had appeared to be two rather similar and parallel bones. The day before, they had been surprised to find that their two bones were actually sides of the same bone. Now, the longer they worked, the closer together they crept, until they were getting into each other's way. Thus, the good-natured shoving.

"I don't know what made me think of it," Poole recalls, "but it suddenly occurred to me that the piece I was working on looked like a deltopectoral crest, and I said, 'Hey, I think this is a humerus!'[1] And Josh said, 'No, you dumbass, if it was a humerus it would go all the way down there,' and he pointed a couple of yards away from where we were working.

"And I said, 'Yeah, you're right, no way. Forget it. But stay out of my light.'"

For two weeks, the paleontologists on the Bahariya Dinosaur Project had repeatedly swung from high to low on sites that started out promising but ended up disappointing. Matt Lamanna wrote in his field book, "I'm worried that we're not making enough discoveries fast enough and that we're losing so many days to bad weather and illness."

Trying to make the best of it, Josh Smith mused, "Look, we're asking a lot of this place. We've never been here before; reconnaissance and prospecting chew up a lot of time wherever you go, and this is our first official season. Even if we come back next year, we'll still just be getting to know this place. Combine that with the environment we're working in and the effect it has on bones, and it's no wonder the quality of what we're finding is so poor.

"And yet there are dinosaurs all around us. We've got probable *Spinosaurus* and *Carcharodontosaurus* teeth. We've got a fibula in one place and what almost certainly are the vertebrae of a large sauropod, perhaps Stromer's *Aegyptosaurus*, in another. And of course we've got amazing fossils from all sorts of other nondinosaur species. What we're missing is that big articulated skeleton."

The fact that Smith and his colleagues had found anything at all was miraculous, something enthusiastic paleontologists often forget in the heat of the moment. Any fossil is a miracle. The odds of one being created in the first place, much less being found, are vanishingly small. Consider the process. First, a creature must die. But not just any creature; it has to be one with a significant number of hard bones. For example, the reason why Stromer found the Bahariya Oasis floor littered with prehistoric shark's teeth and fin spines but nothing else from a shark is because that is virtually all that is bony in a shark. Moreover, the remains of the dead creature have to be sturdy; delicate parts tend to be destroyed over time. This is why there are so few fossil remains of the small, lightly boned early mammals Stromer worked so hard and so unsuccessfully to find; they just get crushed or washed away.

Next—and this is perhaps the trickiest part—the body of the creature must be buried relatively quickly, before the elements and scavengers can break it down and scatter the remains. The best dinosaur remains tend to be found in places where a catastrophe occurred. There is a more recent analogy. In A.D. 79, Italy's Mount Vesuvius erupted violently, heaving into the air prodigious quantities of ash and hot cinders, which, when gravity overcame the force of the eruption, fell upon the city of Pompeii at the mountain's foot. Everything and everyone in the city at that moment was instantly entombed in the cinders and ash, sealed hermetically for all time—or at least until the late eighteenth century, when a German archaeologist began unearthing the city.

Roughly the same kind of thing has to happen to create fossils. Sometimes the process is catastrophic. It might be a massive flood that suddenly inundates an entire valley, burying it in mud of the sort that covered the thousands of dinosaur eggs recently discovered in Patagonia.[2] Or perhaps a stupendous submarine mudslide of the type that resulted in the Burgess Shale of western Canada, which has yielded extraordinary fossils from the Cambrian Period back near the beginning of the Phanerozoic Eon, some 500 million years ago. Or the terrible sandstorms thought by some scientists to have occurred during the Cretaceous Period in what today is Mongolia, burying dinosaurs where they crouched, turned away from the terrible wind. In some cases, the process can be more mundane, as simple and as deadly as a mud pit or quicksand from which a lumbering dinosaur struggled unsuccessfully to free itself. Or an animal might simply perish near the shore of a lake or a sea and slip down to the quiet bottom or became buried in tidal sediments. The burial, however, may not be permanent. Rivers can bury skeletons in sediment and uncover them again repeatedly.

Next, the environment in which the creature is buried needs to be fairly peaceful for long periods of time, so additional layers of sediment can settle atop the body. Sometimes bones can emerge millions of years later virtually unaltered. More often, moisture filtering down through the sediment will pick up minerals and then deposit them again in the interstices of the bone, or replace the bone altogether, a process called permineralization, which effectively turns bone to rock. More time passes and the deposition of new sediments continues. The weight of the successive layers may well flatten the now fossilized bone. Violent activity deep within the Earth's crust may lift, twist, or even bury farther the rock layer in which the skeleton is entombed.

Finally, erosion must strip away all of the layers of rock that have accumulated above the fossilized skeleton. The process of building up must be matched, and eventually exceeded by, the process of tearing down. In time—maybe tens, even hundreds, of millions of years—the forces of erosion, along a stream or riverbed, for example, may cut through the overlying sediments and expose the fossilized bones.

Then someone must wander by. Someone who has the kind of eye that can recognize the difference between old bone and old rock, a rare skill. And that person must have the curiosity and energy to bend down, look hard, begin digging, and release the creature, finally, after millions and millions of years of being locked in rock.

The statistical probability of this precise sequence of events happening is almost but not quite nil, which is one reason why the fossil record is so spotty. Based upon an estimated rate of species longevity, scientists estimate that the fossil record of the last 500 million years should contain the remains of at least 500 million species. To date, only some 500,000 species have been identified—a sample of just one tenth of 1 percent. Of that total, fewer than 500 species of dinosaurs have been identified definitively.[3]

Little wonder, then, that the Bahariya Dinosaur Project team found the search frustrating. The odds were stacked against them.

Ernst Stromer had beaten those odds in Bahariya. Between 1911 and 1914, he and Richard Markgraf had found not one but three gigantic meat-eating dinosaurs. Says Dr. Thomas R. Holtz, noted paleontologist and director of the Earth, Life and Time Program at the University of Maryland, "If it were not for Bahariya and Stromer's discoveries there, we would have no idea even to this day that there could have been three giant carnivores in one place. It just doesn't happen."[4]

Stromer's discoveries created what has come to be known, among those who know about him at all, as "Stromer's three-theropod riddle." The Royal Ontario Museum's Hans-Dieter Sues explains: "In most prehistoric ecosystems, you typically have one top dog, so to speak. For example, in the Late Cretaceous in North America, it was *T. rex*. But in Bahariya you had three huge predators: *Spinosaurus, Carcharodontosaurus,* and *Bahariasaurus,* all living in the same environment. So here is the riddle, which Stromer himself wondered about: What on Earth were they eating?"

Stromer found the remains of dozens of species of plants and animals in Bahariya. Most of the animal remains were of marine creatures, not land-dwelling ones. Though he found sketchy evidence of a diplodocoid sauropod and another much larger sauropod, he was able to identify definitively just one species of plant-eating dinosaur, *Aegyptosaurus,* not an especially large sauropod and unlikely to have been the sole food source for three species of giant meat eaters (and very possibly more, given additional fragmentary evidence described by Stromer). There has been some debate as to whether one of the predators, *Spinosaurus,* might have been a fish eater, but many paleontologists dismiss this notion, concluding it is far more likely that these theropods were opportunistic carnivores, eating whatever meat came their way—including fish.

But Stromer's riddle remains. It baffled Matt Lamanna. "These three predators each weighed three tons or more. I know of no other dinosaurian ecosystem that supported three carnivores this size. They would have required a lot of food to stay alive. The ecosystem that supported them must have been incredibly productive."

"Granted," says Peter Dodson, "as successful as Stromer was, we still don't know a great deal about the diversity of fauna in the environment of Late Cretaceous Bahariya. But there is no question that the fauna we do know about is extraordinarily unbalanced. Under normal circumstances, you expect a large variety of plant eaters, sauropods, feeding a limited number of meat eaters, theropods. But here the situation is reversed; we have more theropods than sauropods. It's definitely a puzzle."

At least in part, the puzzle, Stromer's riddle, was about to be solved.

On January 27, 2000, after days during which several members of the expedition team had been laid low by intestinal unpleasantnesses of varying degrees of severity, the team was back in the field, and project leader Josh Smith made what would turn out to be a pivotal decision.

"We had gone to Gebel el Dist as soon as we arrived in Bahariya because we knew that was the area where Stromer and Markgraf had made many of their best finds. We were sure that if there were dinosaurs to be found they would be found there, in the northern part of the depression. But the bone quality was lousy—badly weathered, shot through with gypsum. We even had bones encrusted with prehistoric snails that had been fossilized right along with the bones themselves. And the rock layers were tough, too; lots of ironstone.

"Folks were worn out, beaten up by the work and the sandstorms, tired and beginning to be unhappy. And we still had three weeks left to go in the season. We needed a change of scene. El Dist was just not panning out."

There was still a great deal of work left to finish up the sites the team had already excavated, and that work continued for several days, until the end of the month—final pedestaling, jacketing, undercutting and flipping the jackets, plastering the bottoms, numbering and transporting dozens of specimens back to the lodge. But in the meantime, Josh took a few members of the team across the desert to the site where he had first come across dinosaur bones during his reconnaissance in 1999. The site, while still in the northern half of the Bahariya Depression, was roughly six miles east southeast from Gebel el Dist, on the other side of the road between Cairo and Bawiti. It was a particularly bleak part of the valley floor, a nearly flat area of desert pavement not far from a low mesa called Gebel el Fagga.

"The rock here was less gypsum-infused and a little more conducive to good preservation, if we were to find anything," says Smith.

Using the coordinates he had noted a year earlier, Smith found the original site and the thick black fragments of bone that had caught his eye from Giegengack's Land Cruiser on that first day in 1999. Here the team split up. Matt Lamanna led graduate student Allison Tumarkin and volunteer Steve Kurth on a prospecting foray, and Smith and Poole went to work on the original location.

"The first thing we did," recalls Poole, "was uncover the site. We just got out some big, stiff bristle brushes, the kind you'd use to scrub a floor, and we basically just bulldozed the sand out of the way. I don't think Josh had any real hope for this site; he thought the bone he'd found in 1999 was just an isolated piece, like so much else we found. But as we moved the sand, we found some other, smaller bones emerging. And we worked a little farther down and found some larger bones. And the larger bones led us to some gigantic bones, and that's when the story started to get really interesting."

At dinner that evening, the mood of the group had definitely changed for the better.

"There was no question we were dealing with much better bone quality at this new site than we'd had at el Dist," says Smith, "and that really helped."

Poole agrees. "We couldn't have asked for better bone quality. Well, we could, but we wouldn't get it. Not out there. So this was great. This was fine. We could deal with this."

One reason the bone was in such good shape was because the rock in which it was encased was, in most places, a dense gray mudstone. This was both good news and bad news. It kept the bones better preserved than sandstone, but it was also very hard. Excavation would not be easy.

And even as they began excavating, no one really had a clue what they were unearthing.

While Josh and Jason had been evaluating this new site, Matt Lamanna and his exploratory group found themselves in trouble with the boss. "We had wandered quite a ways, prospecting, and at one point I saw a group of hills that looked intriguing. I went over there and saw something in a sandstone outcrop that looked just like dinosaur skin. It wasn't skin, as it turned out, but I began to realize there were bluish objects in this outcrop that were quite clearly well-preserved bone. Lots of it. Bits of turtle, lungfish jaws, and other things. A little higher I found the metatarsal of a dinosaur and all kinds of small vertebrae. Then I heard the radio crackle and realized we were out of radio range, and I knew Josh was going to be pissed off."

"Josh is very protective of his crew," Jason explains. "He really gets worried if someone slips out of radio range. He's responsible. He was angry."

Matt goes on, "We'd argued back and forth over the radio as my team walked back, so when we got to where Josh and Jason were working, I said, 'Okay, fine, I'm not going to show you what we found.' And he kind of made a face and said, 'Okay, show me.' So I showed him some lungfish jaws, and he said, 'Nice, well preserved, but I'm not impressed. Whip out something else.' So I pulled out the metatarsal and his eyes just lit up. And that was the end of that. But we never got out of radio range again, either."

The site Matt found—on the date of his brother Jon's birthday and thus known thereafter as the "Jon's Birthday" site—would eventually become the richest site of the entire expedition, in terms of the number of fossils it produced. But it would not be the most dramatic. That distinction was about to be claimed by the site Josh and Jason were working on near Gebel el Fagga, beneath the bone fragments that Josh, Jennifer Smith, and Bob Giegengack had decided in 1999 were "of limited scientific value."

While Matt's group began exploring Jon's Birthday, and Peter's team continued making interesting discoveries back at Lizard Ridge, Josh's group began slowly unearthing bone at the new site. Josh and Jason, helped occasionally by other members of the team, painstakingly cleared sand, gravel, and rock away from the emerging bones with smaller brushes, awls, and dental picks. A remarkable variety of bones began to emerge—large bones, clearly those of a dinosaur. And the bones told Josh they had belonged to a sauropod, a plant eater.

A few days into this process, he called Peter Dodson over from where he was working at Gebel el Dist to examine what they were uncovering. Ever the cautious elder, Dodson crawled around the new site and studied the bones peeking out of the rock. Then he stood up, put his hands on his hips, and said, "You know what I think, guys? I think 'Yahoo!'" For the first time since they'd arrived in the oasis, Peter Dodson was excited. And in a gesture typical of him, he went back to Gebel el Dist and let his students have the excitement of opening up the new site.

The work progressed inch by inch in the hard, blocky mudstone and was made more difficult by the fact that in some places bones overlaid one another. "We could see some ribs in the rock, maybe a scapula or shoulder bone, and a limb bone or two, all nicely preserved," remembers Josh. "It looked to me like a small sauropod, maybe a juvenile version of Stromer's *Aegyptosaurus*." About this, Smith soon would be proved rather dramatically wrong.

By this time, geologists Ken Lacovara and Jennifer Smith had begun to wander farther afield. While exploring Gebel Maghrafa, a long limestone-capped ridge just west of Gebel el Dist in the shadow of the edge of the escarpment, they kept up their record of finding most of the expedition's dinosaur bones, which, in a sense, is not surprising. As Lacovara comments with evident amusement, "We were just covering a lot more territory than the paleontologists were. We'd see something, log it in the GPS, and let the rest of the guys know later. Paleontologists don't cover very much ground. They see something promising, and the next thing you know they've dropped to their knees and gotten out their toothbrushes."

Lacovara found the new bone about a quarter of the way up the slope of the hill, under a fractured layer of ironstone. Jen Smith describes it as "a really distinctive bluish-purple color, because the bone had been completely replaced with iron minerals." There was more bone nearby. But by this time Lamanna's team was busy with Jon's Birthday, Josh's team was hard at work at the new sauropod site near Gebel el Fagga, and Peter Dodson's crew was giving the final touches to Lizard Ridge. They put the intriguing new find, which seemed to Lamanna to have some promise of being an associated skeleton—one disassembled but proximate—on the back burner. Said Jen Smith, "All of a sudden, there was as much work—and good work—as we could do."

By the beginning of February, Peter Dodson, Patti Kane-Vanni, Medhat Said, and Yousry Attia finished up Lizard Ridge and joined the rest of the Bahariya Dinosaur Project team working farther south. Dodson, Kane-Vanni, and Yousry, all of whom were interested in smaller specimens—Yousry, for example, still hoped to find early mammals—joined Matt and the others at Jon's Birthday. Medhat joined Yassir, Jean Caton, Allison Tumarkin, Jason, and Josh at the peculiar pile of large bones now called the South Sauropod site, which by now had been gridded out in two-meter squares. On different days, various members of the teams would swap places for a change of scene, scale, and muscle soreness.

On February 1, 2000, Jason Poole cleaned a new spot at the South Sauropod site and began chipping away the stone matrix around what looked like a thicker bone than the ribs the team had been excavating. "Although it was almost entirely invisible, what I could see looked like it might be something like a radius or an ulna from the forelimb of the animal. After a while, Josh came over and started working near me, and pretty soon he had uncovered a similar bone running parallel to mine. So I was definitely thinking now they were a radius and ulna in place."

Smith recalls, "I thought it was more likely they were two more ribs in close association. But you have to remember, this is a pretty slow procedure; things get revealed very slowly, and you have lots of time to spin out theories to break the tedium. We were working with dental picks, little brushes, and occasionally a hammer and awl, very carefully removing rock from around the bone and trying not to mark the bone in the process.

"We were each chipping our way down the length and sides of our respective bones at maybe an inch or two an hour, and it became clear my bone was larger. I had chipped away the rock on one small section of the top and halfway down each side, and it was looking perfectly cylindrical, with a diameter of maybe four inches. So the rib idea sort of went out the window. Maybe we were working on a limb and an ilium, I didn't know. I remember radioing Peter at the other site toward the end of the day and joking with him that I hadn't found a bone after all; it was just an old stovepipe."

The next day, which dawned clear and warmer, Smith and Poole were back at work, chipping rock alongside each other and trading barbs. Sometime around lunchtime, Jason's bone suddenly became baffling. Instead of continuing around in cylindrical fashion, the side of the bone started to curve downward and outward, toward the bone Josh was working on. It made no sense at all.

"So right away I know I'm not working on a radius or an ulna anymore," he says. "Then Josh, who was working the inside curve of his own bone, found it was doing the same thing. Instead of continuing around like a stovepipe, it was flaring out toward mine. That was our first 'Aha!' moment, when we looked at each other and said, 'You know what? We're working on two parts of the same bone.' It was a complete mystery. Totally bizarre."

Matt Lamanna recalls, "At one point around midday, while I was working with Peter and a few of the others at Jon's Birthday, Josh came on the radio saying, 'Hey, I think you'd better get over here.' So I drove over and looked at what they had begun to uncover, which was this bone exposed about a foot long with some kind of furrow running down the middle. I was at a complete loss. We were sure it was a sauropod, and the only thing I could think of that it might have been was an *Aegyptosaurus* hipbone. So I just left them to it and went back to Jon's Birthday."

By the end of the day, things were no clearer.

On February 3, Josh and Jason continued working on what now was clearly one bone, painstakingly revealing more of it. "It was a really beautiful day," Poole remembers, "really warm and dry, a great day to chip rock." By now they had abandoned the thought that the mysterious bone they were quarrying might be a hipbone and were working on the theory that it was a femur, the largest leg bone. But during the afternoon, Jason's side threw him another curve when the trend of the upper edge started to rise above the level at which the rest of the bone edge had been running. Femora don't do that.

"That was when I had the second 'Aha!' moment," says Poole. "I know only one bone in the body that has this odd sort of protrusion on the side, and that's the upper bone in the arm, the one between the elbow and the shoulder. So I sort of blurted out, 'You know what? I think this might be a humerus!' And Josh just looked at me like I was an idiot. That's when he pointed a couple of yards away and said, 'No, you dumbass; if it was a humerus it would go all the way down there.' "

Thinking back to that moment, Smith grins and says now, "I think in the future I'll just keep my mouth shut when I'm in the field."

A few hours later, as evening approached and they exposed more of the length of the bone, it became clear that it did indeed go "all the way down there."

They radioed Lamanna again. Since it was the end of the day, the entire Jon's Birthday crew arrived. "When they got there, Josh and I were just sitting there with these big grins on our faces," Jason remembers, "and as Matt and Peter walked up, Matt was looking at us and saying 'What? What?' and then we just watched as both of their jaws dropped open."

Uncharacteristically, Lamanna was dumbstruck: "Just six months earlier, Peter, Allison, Jason, and I had uncovered the skeleton of a large fifty-foot sauropod in Montana that included an extremely well-preserved humerus. And it was big, maybe thirty inches long. But this bone was more than double that size. This bone was just gigantic."

"Matt does this thing when he's really thinking hard," says Poole. "He sort of taps or scratches his nose. He probably doesn't even know he's doing it. We gave him our calculations on the bone's measurements, and you could watch him tapping his nose and going through all these numbers and diagrams he keeps in his head, and finally he just said, sort of amazed, 'This does not match anything I have ever seen. This could be a new species. This could be the largest dinosaur in the world.' "

Says Lamanna, simply, "It was just staggering to think about how big that animal was."

"We were all pretty happy," Poole remembers. "After all the digging and dead ends, all the wandering around in the desert and the cold and the sandstorms, I thought to myself, 'Hey, we just paid all the bills!' "

Said project leader Josh Smith later that night, "I think we know now what was keeping all those theropods alive in Bahariya. Maybe we just solved part of Stromer's riddle."

The process by which paleontologists identify and classify dinosaur fossils is called systematics. This is something of a misnomer. Anarchymatics might be a more accurate term. For a start, there are three different and competing approaches to systematics.[5] Paleontologists argue constantly about the various strengths and weaknesses of each approach. They also argue constantly about the relative evolutionary position of dinosaur species. In fact, this suits the science splendidly. The science itself is fluid; it is perhaps appropriate that its classification systems are fluid as well.

The basic concept behind each of these systems is relatively simple, like all good scientific concepts. It was first proposed in the mideighteenth century by Swedish botanist Carl von Linné, more widely known by the rather grander Latin name he chose for himself as a scholar, Carolus Linnaeus. Linnaeus suggested all life could be classified by similarities of visible form into a sort of hierarchy of groups and subgroups, called *taxa* (singular, *taxon*), from most unique (species), through gradually larger groupings (genus, order, class), to most all-encompassing (kingdom—as in "animal kingdom" or "plant kingdom"). Of course, in predating Charles Darwin by almost a century, Linnaeus could provide no clue to how these variations came about. But it was a start. And it created a little crimp in the then prevailing notion of a "Great Chain of Being" that, naturally, put man on top. Darwin's proposition that species evolve through time, combined with Linnaeus's classification scheme, led to the creation of various "trees of life" in the century that followed publication of Darwin's *The Origin of Species*. Ultimately, these mapping exercises also failed, for the simple reason that evolution itself had created so much species diversity over such a long time—most of it already extinct—that there were far too many data points missing (and far too many others that were just confusing) from which to come up with a single coherent diagram. What's more, the whole idea of hierarchy became pointless; there was no reason to believe one animal was any more important than another.

In the mid-twentieth century, an East German entomologist came up with a lucid system that retained Linnaeus's essential determining criterion—living things can be organized based upon their anatomical similarities—but also provided a way to handle the vast gaps in existing knowledge. Willi Hennig's system, which came to be called cladistics (it comes from the Greek word *klados*, or "branch"), takes advantage of the very species diversity that bedeviled earlier efforts at classification. Take any newly discovered species, said Hennig, and look back. Find the most recent ancestor that appeared with a particular evolutionary novelty and you have the spot where two species that share this common ancestral characteristic—their derived characteristic—first diverged. Cladistics is simply the successive branching from common ancestors, in any category of living things. It is a bit like the address on an international business letter, with each line above the last becoming successively more particular: country, state or province, city or town, street address, company name, department in company, name of individual.

There are disputes among paleontologists about the utility of this method for determining evolutionary relationships, as there are with other methods, but one of its great strengths is its ability to enable a relatively quick decision on where a newly discovered sauropod—like the one found by Josh Smith and Jason Poole in Bahariya—fits in the existing branching diagram of sauropod dinosaurs discovered by earlier paleontologists.

The evolutionary relationships of dinosaurs have been mapped on cladistic branching diagrams, called cladograms. But these diagrams are far from complete. Indeed, they probably never will be, as most new dinosaur discoveries alter them. Which raises an intriguing question: Fewer than five hundred species of dinosaurs have been identified definitively,[6] but how many were there altogether? A lot more, without question. Peter Dodson explains: "The age of dinosaurs lasted 165 million years, from the Late Triassic to the end of the Cretaceous Period of the Mesozoic Era. It's been estimated that the longevity of a given genus of dinosaur before it died out or evolved into something else was between 5 and 10.5 million years, say an average of 7.7 million years. At any one time there were probably no more than 100 dinosaur genera alive. That suggests that there may have been anywhere from 900 to 1,200 genera of dinosaurs over the course of that 165 million years. Other researchers, using different methodologies, have estimated as many as 3,300. So our classification systems still have more blank spaces than filled spaces."[7]

While Matt Lamanna was standing on the edge of the South Sauropod quarry in Bahariya on the afternoon of February 3, after Smith and Poole revealed the gigantic humerus they were excavating, it was precisely this kind of anatomical analysis he was doing in his head. That night Lamanna hit the books he'd brought with him on the expedition.

"There are some kinds of sauropods, called brachiosaurids, that have longer forelimbs than hind limbs," explains Lamanna, "but I didn't think this was one of them—in part because they were probably extinct by then. At the same time, it didn't seem to be a specimen of Stromer's *Aegyptosaurus,* either. I had Stromer's monographs with me, with their beautifully detailed illustrations, and it was unlikely this bone belonged to that animal."

As the excavation proceeded, members of the team began finding other parts of the skeleton: more ribs, the shoulder girdle, a second (ultimately partial) humerus—all huge. Like Lamanna, Josh Smith was still puzzling over the humerus: "A lot of dinosaur identification is about checking out variances in shape. It's comparative anatomy. This new humerus was different from any one we had ever seen before. It was not shaped like a brachiosaurid. It didn't look like an *Aegyptosaurus* humerus, which is boxier and a whole lot shorter. And in Bahariya, if it's not *Aegyptosaurus* or some other taxon from the Late Cretaceous of Africa, like Morocco or Niger, then it's almost certainly a new genus or species."

"You can learn a lot from a humerus," bone preparator Jason Poole explains, "and especially from that bump on one side called the deltopectoral crest. You can look at a dozen species of dinosaurs, and the deltopectoral crest will always be positioned a little differently. It exists on the humerus so muscles can attach the limb to the chest and help move the leg around. And since every animal moves slightly differently, every deltopectoral crest differs. We can begin to figure out what an animal is by asking questions like: How large is the crest compared with the rest of the bone? How far does it stick out from the main shaft of the bone? What is its position on the bone; how far down is it from the top? Are there any unusual concavities—scooped-out places? Each measurement helps us narrow down the list of candidates."

Lamanna was somewhat surprised by the shape of the humerus. "This humerus was remarkably gracile: It started out wide and massive and then tapered toward the center fairly narrowly before broadening out again at the base. I had seen similar but not identical humeri from South America and Utah. We needed more diagnostic evidence. You'd love to find a skull, but you almost never find skulls with sauropods. They're tiny to begin with, given how big these animals were, but they break down quickly after the beast dies. Vertebrae are the next best thing."

Explains Peter Dodson, "Ideally, you hope to find a skeleton that's complete from tip to toe, but we rarely do. So you hope for what are called diagnostic bones, which are bones that vary in some significant way from species to species. A humerus is good. A scapula is good. But you hope for vertebrae from the neck or the dorsal or rear part of the torso and tail."

Soon his wish began to come true. With many members of the team working on the site in the days that followed the discovery of the immense humerus, a few tail vertebrae finally emerged from the quarry. Says Jason Poole, "We got lucky. We pulled out some caudal vertebrae, the spinal bones close to hips at the point where the creature's long tail attached to its torso. And what looked like a cervical vertebra, a neck bone. Though these bones were somewhat less well preserved than others at the site, they had distinctive processes, little arms on the vertebra to which the muscles attach and which direct how the spine will move, that clearly were diagnostic."

Lamanna adds, "I was looking for derived characteristics so I could place these bones in context. We needed to know what characteristics this animal shared with other sauropods of this period in order to identify it. The caudal vertebrae were the giveaway; these vertebrae had a kind of ball-and-socket joint structure, a cup shape on one side into which the ball of the next vertebra fit. That's how their tails articulated. That is almost exclusively a characteristic of titanosaurids."

"There were other clues as well," says Josh Smith. "One end of the humerus was squared off on one side. Some of the processes, the spikes, of the vertebrae were downswept. Like the ball-and-socket structure, these were all anatomical signals that told us this was a titanosaurid."

Titanosaurs were the dominant sauropods of the Late Cretaceous, before the great mass extinction that ended the Mesozoic Era. They have been found in many parts of the world, but most notably in South America, where a number of different genera have been identified. They ranged from elephant-size to earthshakingly enormous.

The problem with titanosaurs is that their fossil record, while widespread, is still fragmentary. No complete articulated skeleton has ever been described. (Indeed, complete, fully articulated skeletons of any dinosaur are rare; most are reconstructed from partial skeletons.) Based upon a few recently discovered bones, scientists had estimated that the largest titanosaur, and the heaviest dinosaur ever known, was *Argentinosaurus,* found in South America. It probably weighed between eighty and a hundred tons, and it may have measured as long as a hundred feet. No humerus had ever been found for it, however. The humerus Smith and Poole were excavating was immense: five feet seven inches long. This new dinosaur emerging from the rock on the floor of the Bahariya Oasis at the foot of Gebel el Fagga would rival *Argentinosaurus* for the title of the most massive animal ever to have walked the Earth.

But the size of this new sauropod was not the only thing unusual about it. Perhaps more unusual was the fact that it existed *at all* in the Late Cretaceous of North Africa.

Not long after he began publishing scientific monographs about the extraordinarily unusual dinosaurs he had found in Bahariya, Ernst Stromer was contacted by the respected American paleontologist William Diller Matthew. Matthew had read Stromer's papers and was impressed by how different Stromer's African dinosaurs were from those discovered in North America.

None of this should have surprised Matthew. After all, there was an ocean between the two continents. Of course the dinosaurs would be different. On the other hand, there was also an ocean between North America and Europe, and there were many similarities between the Early Cretaceous dinosaurs found in these two places. It was a puzzlement. And it wasn't the only one. Matthew certainly knew that the fauna of the Late Triassic and Early Jurassic periods of the Mesozoic— that is, the period from 230 to 180 million years ago—were generally similar across the globe. Now here was Stromer with Late Cretaceous dinosaurs that were utterly different from the ones Matthew was familiar with from roughly that same period in North America. What could have caused dinosaurs that were essentially similar worldwide to have become so very different by the Late Cretaceous?

There was no way for either Matthew or Stromer to have guessed that the continents had ever existed in any part of the globe other than the places they are now located. The notion that the continents were adrift would not be proposed for another decade and would not be proved for half a century. Similarly, there was no way for them to know that North America and Western Europe had once been part of the same landmass as Africa and, more important, had remained connected to each other long after Africa and the rest of the continents then in the Southern Hemisphere had split away.

But it was also the case that the search for dinosaurs in the so-called dark continent was in its infancy (in many respects, it still is). Perhaps William Matthew thought that, in time, more familiar Late Cretaceous dinosaurs would be found in Africa as well. If he did, he was wrong. We now know that fauna were similar across the globe in the Late Triassic and Early Jurassic periods because there was only one continent at that time, Pangaea, and animals were—climate, environment, and diet permitting—free to wander wherever they wished. We know also that they were different by the Late Cretaceous because Pangaea had by then split into the fragments that would form the continents spanning the globe today. Each time the earth split, the species on either side of the rift began to diverge as well.

It is hard to overestimate the importance of Ernst Stromer's discoveries or, for that matter, the tragedy of their loss. Says the University of Maryland's Thomas R. Holtz, "When it was first found, the Bahariya assemblage was the only good collection of specimens of Cenomanian [the first age of the Late Cretaceous] fauna from anywhere in the *world*, not just Africa or the Southern Hemisphere. Then it was lost."[8]

"To oversimplify somewhat," explains Matt Lamanna, "by the Late Cretaceous there were very different faunas in the Northern and Southern hemispheres. The creatures filling various ecological niches were evolutionarily different. To a great extent, except for a *Carcharodon-*

tosaurus skull found recently in Morocco and a few others,[9] Stromer's dinosaurs are still unique; dinosaurs from the Late Cretaceous in Africa are almost totally unknown. *Spinosaurus,* for example, remains virtually unknown. *Aegyptosaurus* remains have never been identified with certainty since Stromer. No one has identified more than isolated remains of *Bahariasaurus.* All we have now are Stromer's papers and their superbly detailed illustrations. But it is clear from these papers that there are few dinosaurs comparable to the ones from Bahariya anywhere in North America, Europe, or Asia.

"Stromer's *Carcharodontosaurus,* for example, looks a lot like *Tyrannosaurus,*" Lamanna continues. "They're both tall, long-limbed, two-legged predators with huge heads, sharp teeth, and small forelimbs. But they're only distantly related. *Carcharodontosaurus*'s closest relative is the theropod *Gigantosaurus,* from South America. We now think we know why that makes sense; though there is still some controversy about the precise sequence, it seems likely that South America was the last landmass to separate from Africa. For that reason alone, I'm not surprised that the only sauropod as massive as the one we've found in Bahariya is South America's *Argentinosaurus.* It just makes sense."

Also for that reason, the gigantic sauropod emerging from the stubborn rock on the undulating desert floor near Gebel el Fagga was a valuable indicator for geologists and paleontologists trying to understand more accurately the timing of the rift between these two great continents, continents whose shapes even today conform to each other's outlines like lovers parting from an embrace.

Getting the sauropod to part from the rock of Bahariya, however, was proving no easy task.

Explains Josh Smith, "The day we found the sauropod was just about the last day we could find anything large and expect to get it out of the ground and prepared for transportation before our time in the oasis was scheduled to run out. We hadn't even identified all the parts in the quarry yet. It was clear it was going to be a brutal schedule."

In the meantime, the other southern site, Jon's Birthday, was continuing to produce a steady stream of intriguing small specimens, the kind of fossils that Josh knew would be important clues to the kind of world the big sauropod lived in and would provide what Peter Dodson called the "supporting characters" in that world. Josh Smith is a relatively unusual paleontologist, one as interested in the environmental context of an ancient landscape as he is in finding dramatic dinosaur fossils. He and Dodson agreed that it was important to keep some of the team at Jon's Birthday, and Dodson headed up that task. Matt Lamanna worked at both sites, and Josh Smith and Jason Poole oversaw the difficult and ultimately dangerous job of excavating, chemically stabilizing, jacketing, and moving the enormous bones that were emerging from the sauropod quarry.

Most of the rock the sauropod bones were encased in, as geologist Jen Smith recalls, was a brutally tough mudstone. "By the last week Ken Lacovara and I were both helping out at the site, and the rock was miserable. We were using hammers and chisels and awls to break out blocks of this rock, and it would come away in pieces so sharp they would cut your hand. The trick is to split away the blocks without breaking the bone as well. Anyone who had any field experience at all was pressed into service."

Meanwhile, the team hired two residents in nearby Bawiti to dig trenches out beyond the bones that had already been excavated to more clearly delineate the extent of the bone bed. The men from Bawiti were so eager to participate that the team members felt the price they were asking for their labor was too low. As Peter Dodson puts it, "We had to bargain mightily for their services; they wanted ten Egyptian pounds a day, but we managed to bargain them up to twenty pounds, and then twenty-five. But we still felt it was a pittance."

As the excavation continued, Dodson's team, having closed down the Jon's Birthday site for the season, joined the rest of the crew excavating the sauropod. Gradually, it became clear that the team would not find a complete skeleton. The site was full of mysteries. The bones lay in sometimes confusing patterns—not random, but not in expected ways, either. And there were large areas of the site that held nothing at all. There were no geologic signs that the skeleton had been broken up by water; the rock it was found in was the kind laid down under quiet environmental conditions. Then Matt Lamanna found a possible *Carcharodontosaurus* tooth mixed in with some of the bones. Another digger found what may have been tooth marks on one of the bones. A possible picture of the great beast's last moments began to form.

"It seems possible," Josh Smith explains, "that this critter was scavenged. The tooth we found belonged to a very large predator, perhaps five tons in weight. It might well have had the capability to completely disarticulate the body, just tear it to pieces and spread it around the site pretty widely. If there was more than one predator at work, which is certainly possible, it's entirely likely each dragged a piece off a few dozen yards to eat in peace. We suspected there was more to our sauropod very nearby, but we had run out of time."

As the days left grew fewer, the pace of the excavation grew more feverish and the hours the team spent in the quarry longer. "Toward the end," says Ken Lacovara, laughing, "we were pulling bones out of the ground in the dark, using the headlights of our car to see what we were doing."

"Pulling bones out of the ground" is more than a little euphemistic. What the team was doing—partly because of the sheer size of the individual bone specimens and partly because in some cases the bones could not be safely separated in the field—was excavating and jacketing huge blocks of rock and bone, some of them weighing in excess of eight hundred pounds. This presented a problem of equally prodigious proportions: how to flip the top-jacketed blocks so the bottom could also be reinforced and plastered.

"We'd all gather around one of these things," remembers Jason Poole, "count to three, and lift. Now, we are pretty buff guys. Ken and Matt lift weights, and Josh is built like a tank. And it would be, like, no movement. Forget it." On a couple of occasions, they tied rope around the blocks and turned them over as gently as possible with one of the Land Cruisers.

Once the bottom was plastered, the specimens were even heavier and another, more critical challenge presented itself: getting the huge, now fully jacketed specimens off the ground and onto a flatbed truck the team had hired in Bawiti for the trip back to Cairo. A forklift was found in the town, but it was disassembled and it could not be reassembled in time. Then the team's ever enterprising Egyptian colleagues found a tripod and hoist in a nearby village. Ken Lacovara, who had engineering as well as geology training, went to the village to look at it.

"It was aluminum and held together with three rusty bolts that were about a quarter of an inch in diameter. There was a chain hoist attached, like the kind you use to pull engines out of cars. I was skeptical, but we didn't have a lot of choice."

They set up the tripod at the rear of the truck and looped rope and chain around the first of the giant blocks. Wisely, Jennifer Smith, Allison Tumarkin, and Jean Caton stood aside. "You can just imagine the scene," Jen says, laughing. "Here are all these guys, maybe fifteen or twenty of them, testosterone pumping, yelling back and forth in two different languages, each certain that he knows the best way to get this job done. Some are pulling on the chain hoist, some are guiding the block. It was scary."

Adds Lacovara, "It was like trying to hoist a grand piano off the desert floor with a little block and tackle. I was sure those rusty bolts were going to snap at any second and just kill us all."

They didn't. Rickety as the tripod was, the team managed, by nine-thirty P.M. on their last night in Bahariya, to load the final jacketed sauropod block onto the flatbed. Naturally, the heavily loaded truck bogged down in the sand on the way back to the road, but the team at last made it back to El Beshmo Lodge for a celebratory dinner and a very short night: They had an appointment with the director of the Egyptian Geological Survey and Mining Authority in Cairo at nine the following morning. The truck pulled out of the lodge parking lot at three-thirty A.M. on February 19. The Land Cruisers followed at five A.M.

As they zigzagged up the northern escarpment of the Bahariya Oasis and headed northeast toward the rising desert sun, the Bahariya

Dinosaur Project team knew they had gathered much more than a truckload of dinosaur fossils. While the paleontologists had been hammering away to unlock the giant sauropod's bones from rock, geologists Ken Lacovara and Jen Smith had succeeded in unlocking the secret of what the world the huge creature lived in 95 million years ago had looked like.

And that would prove just as remarkable as the dinosaur itself.

TEN

LOST WORLD OF THE LOST DINOSAURS

Ken Lacovara does not recall precisely when the question began insinuating itself into his head. Sometime toward the end of the first week of the expedition, the geologist thinks, but it is unclear. It arrived slowly, the way a worry sometimes does, and then nagged at him with growing insistence until it became intolerable.

It was a problem having to do with sand.

Lacovara likens science to detective work. Not surprisingly, he is fond of quoting Arthur Conan Doyle's great fictional detective, Sherlock Holmes. His favorite is Holmes's definition of deductive reasoning. Says Lacovara, "When you have eliminated what cannot be, whatever is left, no matter how improbable, is what *must* be."

"You never prove anything in scientific research," Lacovara explains. "The best you can do is propose a theory that fails to be disproved. Newton's law of gravity, for example, has never been proved and never will be. It's just failed, so far at least, to be disproved. So, like Holmes, a scientist's job is to eliminate everything that cannot be and reveal the thing that must be."

But there is another Sherlock Holmes passage that, had he recalled it at the time, might have helped Lacovara with his problem having to do with sand. In one Holmes story, the detective and his estimable colleague, Dr. Watson, are discussing a crime that occurred during the previous night at a house where they were both staying. Holmes notes that the most significant issue in the case is "the problem of the dog barking in the night." Watson, the perpetually baffled everyman, replies that there was no dog barking in the night. To which Holmes counters, "Yes, that is the problem."

Ken Lacovara's problem having to do with sand was that there was none. Yes, he had found beautiful deposits of cross-bedded beach sands and even dunes in places on el Dist, but in one particular section, there was no evidence of Late Cretaceous beach sand where there should have been.

Fellow geologist Jen Smith explains: "We had found a layer of rock—a hard, dense, reddish iron oxide–rich rock—the kind that is formed in vegetated tidal flats. It was sitting right on top of a layer of rock that quite clearly was formed in the shallows of a near-shore marine environment. That doesn't happen. Normally, you'd expect a layer of beach sand in between. But it wasn't there. Where was the missing rock? Where was the beach sand?

"I'm a fluvial geologist, which means I know river systems best, and in ancient river systems, you can get all sorts of different sediment combinations and it's no big deal. I'm also a big believer in the theory that, due to erosion, most of the sedimentary record is missing anyway. So when Ken found this one rock layer that, logically, should not have existed on top of the layer immediately below it, I just assumed the intervening layer or layers had simply eroded away. We call it an unconformity. But Ken is a coastal sedimentology expert, and I could tell he didn't feel comfortable with erosion as the explanation for the missing layer. It bugged him."

Lacovara explains: "Oftentimes I read professional papers that describe one depositional environment sitting on top of another that simply cannot exist in that order, and it's dismissed as an unconformity. It drives me nuts; I hate dealing with geological non sequiturs. Calling anything that doesn't belong someplace an unconformity can sometimes just be a copout. We knew that these rocks in the Bahariya Formation had been created in a coastal environment. I knew the beach sand should have been there, and I just didn't see erosion signs. I knew there had to be a better explanation, but I just hadn't figured it out yet."

Ninety-five million years ago, at the beginning of the Late Creta-ceous, the Earth was undergoing a series of profound geological and climatological changes. Global temperatures are believed to have been much higher than they are today, as were carbon dioxide levels. Sea levels were higher than at any time in the last 250 million years. One reason was that because of higher temperatures worldwide, the sea-water was warmer, and because it was warmer, it expanded. What's more, it appears there were no polar ice sheets. But the larger reason for the record-high sea level arose directly from the breakup of the supercontinent of Pangaea, which by the Late Cretaceous was nearly complete.

The fragmentation of Pangaea into the continents we recognize today occurred because the Earth's crust was spreading along volcanic rifts from which hot magma flowed episodically. As the continents split apart, seawater filled the gaps, eventually forming oceans. But the vol-canism continued, and as new magma billowed out of the gashes in the seafloor, this molten rock piled up and cooled, creating massive under-water ridges running thousands of miles parallel to each rift and rising as high as modern terrestrial mountain ranges. As millions of years passed, ridge after ridge was created and then pushed sideways from the edges of these rifts as the next ridge was born, until the bottom of the oceans in many places looked like corduroy and the continents had been pushed thousands of miles apart. The new rock in these immense ridges was warmer and less dense than the rock built by earlier ridges and therefore sat higher on the Earth's mantle. In time, the combination of these phenomena—warm, wet climate; warm seawater expansion; lack of polar ice caps; underwater ridge creation; new, less dense oceanic crust rock—displaced enormous quantities of seawater and caused sea levels to rise hundreds of feet higher than they are today, inundating vast areas of the land with shallow seas. All of what today is central

North America, for example, was covered by one of these seas (called epicontinental, or epeiric, seaways).

The situation was the same everywhere on the globe, and what is North Africa today was no exception. The east-west Tethys Sea was nearing its widest stage of development at this time, and its southern shore lay deep within what today is the Sahara. Indeed, as the Late Cretaceous continued, the western or Moroccan bulge of Africa was cut off completely from the rest of the continent by the joining of the Tethys Sea and the South Atlantic Ocean and stayed that way for millions of years until the seas once again receded.

These epeiric seas were different from oceans in a number of ways. They were narrower, so wind had less space to build waves. The tidal range was mild because many of these seas were somewhat constricted, so the tidal ebb was trying to escape even as the flow of the new tide was coming in, effectively creating a stalemate. In addition, the gradient of the floor of these seas was very gradual, and consequently, storm waves "tripped" and were dissipated long before they ever reached the shore. Finally, North Africa was quite close to the equator, or the Doldrums, an area of the globe where atmospheric winds are very light. The result? Very placid waters lapping gently on the northern shore of Africa. A quiet environment for sediments to be deposited century after century.

There were major changes under way on land as well. Until the beginning of the Cretaceous Period, the plant life that covered the early continents was surprisingly limited and monotonous. Grass did not exist. There were no flowering plants of any kind. As Peter Dodson has pointed out, Mesozoic Era plant life before the Cretaceous Period was limited primarily to conifer and ginkgo trees, cycads, horsetails, and ferns. Lots of ferns.[1] In fact, some of the herbivores of this period had teeth that appear to be shaped explicitly to "comb" foliage off the stems of ferns and conifers.

Even for herbivores this was not especially nutritional stuff, and big plant eaters like the giant sauropods had to consume huge quantities just to stay alive. Scientists believe they evolved enormous guts, and because they lacked the kind of jaw mechanism that would permit chewing, they swallowed the shrubbery whole—and wholesale, eating perhaps a quarter ton of foliage every day.[2] Because of their long necks, it was thought that they ate like giraffes, grazing the tops of trees. But this now appears unlikely. They appear to have had neither the musculature nor the skeletal capability to lift or hold their necks in an elevated position for very long. Indeed, doing so probably would have been fatal. The blood pressure necessary to get blood to a dinosaur's head suddenly lifted to such great heights would have placed its cardiovascular system under unsustainable stress.[3]

It seems far more likely, particularly given how much biomass these great beasts had to consume, that their elongated necks permitted them to strip huge arcs of the landscape at low or moderate elevations without even moving their feet—a far more efficient system. There is some evidence to suggest that sauropods were gregarious—not unlike elephants, feeding and migrating in herds—so the damage they caused to the slow-growing plant life they fed upon must have been astonishing. Trampling the earth, stripping the vegetation, they were huge, four-legged, traveling environmental minidisasters.

The torn-up earth and naked open spaces, combined with the warm, humid environment that prevailed in much of the world at the time, provided an ideal environment for the emergence of the first flowering plants, or angiosperms, in the Early Cretaceous. Angiosperms had a number of distinct advantages. Fast-growing and weedy, they were able quickly to colonize earth disturbed by grazing sauropods. Fast growth also meant they went to seed readily, and rebounded easily after being grazed. By the Late Cretaceous they had begun to create an increasingly diversified array of scrubby plants and even trees in many parts of the globe.[4]

It was the angiosperms that made it possible for a large and diverse group of herbivores—the ornithopods—to proliferate during the Cretaceous Period, roaming the landscape, especially in North America and Asia, in herds. Generally smaller than the huge sauropod herbivores of the Jurassic and Early Cretaceous and typically equipped with teeth for grinding and chewing fodder, they were, says the Royal Ontario Museum's Hans-Dieter Sues, the "cattle of the Cretaceous." By the Late Cretaceous, sauropods had begun to decline in some parts of the world, especially North America and Asia, but they appear to have remained the dominant herbivore in Africa. The sauropod discovered by the Bahariya Dinosaur Project was a prime example. Indeed, neither Stromer and Markgraf nor the 2000 expedition team ever found evidence of ornithopods, though they have been found in Early Cretaceous sediments in Niger.

This is not to say that there were none. The fact is, there has been so little exploration of the Late Cretaceous paleontology of Africa that not much is known about what life was like on the continent then. While the period is thought to have been hot and wet globally, there are many questions about the precise extent of those lush, humid conditions. Some scientists suggest that the northern fringes of southern continents, and Africa in particular, had persistently hot, arid conditions.[5]

But if that was the case, it puts a new twist on Stromer's riddle. If there were three huge meat-eating dinosaurs roaming the northern coastal region of Late Cretaceous Africa and at least two species of large plant-eating dinosaurs in sufficient quantity to be part of the carnivores' diet, what were the herbivores eating if the landscape was arid? Ken Lacovara's Holmesian habit of deductive reasoning told him it didn't add up. The region of Africa where the Bahariya Oasis now sits had to have been prodigiously productive, capable of creating huge quantities of biomass to feed massive plant eaters, which in turn fed terrifyingly large meat eaters. But if it was arid, what were the herbivores eating?

Lacovara was about to discover that the answer to this question might well be found in the answer to the other question that vexed him: the problem of the missing beach sand.

While Ken Lacovara and Jen Smith continued to measure, map, and document the successive layers from the bottom of Gebel el Dist to the top, Lacovara kept coming back to the one part of the section that troubled him—a section low on the hill, close to the level where his colleagues were finding dinosaurs.

"In this particular spot," explains Lacovara, "we had one layer of glauconitic rock. Glauconite is a mineral that often precipitates out of shallow seawater and settles into marine sediment. It's sort of greenish and easy to recognize. You also find a lot of fish scales, marine snails, shells of bivalves like oysters in this kind of rock—all the things you find in shallow coastal water, say one to twenty feet deep. But sitting directly on top of that layer, effectively welded to it, was a layer of rock consistent with a vegetated tidal flat. It shouldn't have been there.

"There is a very important principle in geology called Walther's Law, that says that in any uninterrupted sequence of sedimentary deposits—the kind we were seeing on el Dist—the lateral sequence of sediments that existed at the time of formation should be mimicked by the vertical sequence of rock layers when we expose them today. We knew that ninety-five million years ago or so, this had been the southern shore of the Tethys Sea. The sequence we would expect to see, therefore, would be nearshore sands with glauconite, then beach or barrier island sand, then a lagoon with vegetated tidal flats, and finally the mainland. When the sea is advancing, as it was during this period, one layer sort of rolls over and covers the next, so you get a nice vertical record.

"But the beach or barrier island sand was missing. Not only that, but the sequence of what was there was upside down. There were intertidal, or lagoon, deposits sitting atop shallow glauconitic marine or shore-face deposits. In a normal situation with an advancing sea, it should have been the other way around. This was evidence that the opposite was true: The land had advanced into the sea. Even so, there should have been evidence of the passing of the beach. That evidence is usually found in the form of a chewed-up gravel layer, full of shells and other debris. This layer is formed where waves break. Their churning action winnows out all but the coarsest material, leaving behind an easy-to-recognize 'lag.' I kept going back to look at the two layers, but there was no lag there. I finally accepted that I was seeing a perfectly gradual transition from shallow marine environment to tidal flat. There never was an intervening beach."

Ken Lacovara is a reasonably normal fellow. But he has one peculiar quirk. He hates to sleep. "My wife thinks I'm crazy," he confesses. "But sleep just annoys me; it's such a waste of time. Once I go to sleep, I know I'll never get that day back. It's gone." Even when he is tired, Lacovara doesn't sleep well and, as a consequence, has plenty of time to think.

"Sometime in the beginning of our second week in Bahariya," says Lacovara, "I was lying in bed in the dark in the room I was sharing with Matt, wrestling with this problem of the missing beach sand, when I realized I couldn't solve it without clearing away all my assumptions and starting over."

He began with a deceptively simple question: What did Bahariya look like 99 million years ago? He put himself on the shore of the Tethys Sea and, in his head, looked around. The water was very calm; he knew that from the grain size of the sediments he and Jen Smith had uncovered. He also knew that the sea level was rising, rather than falling, because there was no evidence in the el Dist sediments of an erosional environment, of rivers flowing downhill toward a receding seashore. He knew he was seeing what coastal geologists call a transgressive sequence.

"Even if we hadn't had the sedimentary records," Lacovara comments, "we might have been able to deduce that from the fact that we were finding dinosaur bones. You find dinosaur bones in depositional environments that have eroded—not, for the most part, in environments that were erosional in the first place."

On this ancient shoreline, he knew there were some places where there were beaches, because he and Smith had found beautifully cross-bedded sandstones characteristic of beach deposits. And in other places at the same general section of el Dist, they had seen evidence of tidal channels and current movement in deposits with what is known as flaser bedding.

"If you could put your head underwater while the tide is gradually receding through a channel," Lacovara explains, "you'd see ripples on the sandy bottom. Then, at the moment of slack tide, you'd see finer silty sediments settling into the hollows between those ripples. When the tide came in again, you'd see the bottom sand creeping back over these little mud deposits, gently covering them. This sequence repeats itself four times a day, with each tidal change. If the sediments remain undisturbed, eventually the record of each change in the tide in that channel is preserved for all time by rock, with little lenses of siltstone encased in the surrounding sandstone. That's flaser bedding. It proves bidirectional water flow existed and that the current wasn't terribly strong. We also found deposits that had what is known as wavy bedding, which are about half sand and half mud, where the movement is even gentler."

And finally, there were those areas of troublesome unconformity, where intertidal sediments—called paralic sediments—were sitting atop nearshore marine sediments. Normally, when sea levels rise, marine sediments will cover over beach and intertidal or lagoon areas, as the seafloor gradually migrates landward; that would be a conformable sequence. But it seemed to Lacovara that instead of being buried by marine sediments as the sea level rose, the intertidal sediments in what he and Jennifer Smith were calling the unconformable sequence were doing the opposite: actually advancing into the marine zone.

"But they weren't unconformable sequences," says Lacovara. "In the case of an advancing sea, the typical sequence is shallow marine sediments, over beach, over tidal flat. A retreating sea normally leaves beach over shallow marine sediments. But here we had neither. We had tidal flat over shallow marine. Nature was clear but we were not. We needed a new paradigm, one where this sequence would be conformable, where it would make sense."

In the dark, Lacovara tried to imagine a coastal landscape in which all of these features might coexist. And later that night, he realized he could. In fact, he did not even have to imagine this landscape. He had been there.

On the Gulf coast of the Florida peninsula, at the western or seaward edge of the Everglades between Cape Sable and Cape Romano, is a vast region the Florida tourism office calls the Ten Thousand Islands. It is a watery world of small, densely forested islands separated by a maze of sandy tidal channels. But they are not islands in the conventional sense, sandy hills rising above the surface of the sea and colonized by plants. These islands are thickets of treelike plants that grow directly out of the seawater, their roots anchored in shallow marine sediments. They are mangrove islands. On the seaward margin, some of these islands are connected by stretches of beach where the sand is finely grained, as the Gulf is normally calm. In other places, the islands have advanced out into the shallow waters of the continental shelf.

Four years before the Bahariya Dinosaur Project went to Egypt, Ken Lacovara spent a brief vacation kayaking in the Ten Thousand Islands. Remembering that trip as he lay in bed at El Beshmo Lodge, Lacovara realized he had his answer at last: 99 million years ago, the southern coast of the Tethys Sea had been an intertidal forest—a mangrove swamp.

The word "mangrove" has several definitions. The narrowest is any of a variety of modern tropical shrubs and trees that grow in dense thickets on stiltlike support roots along quiet tidal shorelines. A broader definition refers to any collection of woody plants that have the same habit of growing with their "feet" in salt water. All have a number of common characteristics. They have filtering systems by which to extract fresh water from salt water. They support themselves on thick "prop" roots that anchor them to the bottom. They connect with one another through a tangled mass of horizontal roots, forming small islands that, in time, can become larger islands as dead leaves and other trapped organic matter settles between the roots, forming a peaty organic soil. Moreover, they often advance seaward out across the shallow marine shelf, forming new islands and thus expanding the coastal margin even if the sea level has been rising. On Florida's Gulf coast, for example, the sea level has been rising gently for eighteen thousand years, but the Ten Thousand Islands mangrove environment has been getting wider, not narrower.[6]

Says Ken Lacovara, "It's often the case in science that the correct answer to any question is simpler than the incorrect answer. There is a sort of informal principle of parsimony that cuts through convoluted explanations. It's called Occam's Razor, and it was first proposed by a fourteenth-century English friar and thinker, William of Occam. He said, and I'm paraphrasing here from the Latin, that one should not increase, beyond what is necessary, the entities required to explain something. Which is to say, don't make things more complicated than they have to be. Nothing demonstrates this principle better than how vastly more simple our understanding of the solar system became after Copernicus suggested that the Earth and all the other planets revolved around the sun, instead of everything revolving around the Earth. The Earth-centered model required the sun and the other planets to do all sorts of complex and, frankly, physically impossible things, while the Copernican model of concentric rings around the sun was elegantly simple."

Lacovara's mangrove hypothesis was another example of the inherent simplicity of the right answer.

The next day, Lacovara ran his idea by Jen Smith. "We were walking by the unconformity," Smith recalls, "and Ken just said, 'You know what this might be?' and presented his hypothesis. I remember thinking, 'Wow, that is exactly what we are seeing here. Suddenly everything makes sense.' Things just started falling into place."

In fact, once he returned to Philadelphia and began digging into the literature, Lacovara would find support for his idea. Another researcher working in the southern reaches of the Bahariya Oasis some years earlier had noticed "massive root colonies [that] resemble mangrove swamp type deposits."[7]

But Lacovara's notion that the floor and lower levels of the Bahariya Depression had been an intertidal tropical-forest environment did not solve just the problem of the missing beach sand—the alleged "unconformity"—between beds of nearshore marine sediment and lagoonal deposits. It solved other problems as well. The sedimentary rock in this area—the same area from which their paleontological colleagues had been removing dinosaur and other bones—had revealed itself to be complex, discontinuous, and puzzling. A perfectly preserved deposit of dune sand would, for example, suddenly end and give way to a dense, hard brown mudstone. This, in turn, might taper off in a short distance to a grainy sandstone with flaser bedding. Fossilized wood stems might sit within a few inches of fossilized oyster-shell beds.

Like the sun burning through fog, a clear picture began to emerge from these otherwise puzzling deposits. "The lateral variability of Bahariya deposits was bizarre," says Lacovara. "In most coastal environments in the world, if you hit a layer of sandstone, for example, it can run for miles, even hundreds of miles. But in Bahariya, when you hit a layer of sandstone, it might go only for twenty feet. And then it would turn into mudstone, and then interbedded sand and mud. It was a mess.

"But in a place like the Ten Thousand Islands in Florida, you can step off the beach and in five feet be in marine sediment. Step five feet in another direction and you're in mangrove deposits, or in intertidal sediments. It's exactly the crazy lateral variability we see in Bahariya."

Lacovara postulated that the hillocks of brown, iron-rich mudstone atop grayish-green glauconitic marine sands were mangrove islands, also known as mangals. Eventually, Lacovara would find more and more evidence of roots welding these two layers together, just as they do in Florida's Ten Thousand Islands, as well as thicker support stems. The altered ancient soils, or paleosols, formed within these islands are different from other forms of sedimentary rock. There are no clear stratifications, no obvious soil horizons. The layers have been disturbed by worms, clams, and other marine life burrowing through them. In addition, the saltwater content of such soil makes them bad places for the preservation of plant material, though bits and pieces are scattered throughout the rock. Some of the pieces even show signs of having been turned to charcoal; brush and forest fires are common in such environments, even in the Everglades.

In between the ancient mangals, the geologists had found grainy linear quartz-sand deposits with flaser bedding. But these deposits were very narrow. Now they knew why: They were the sandy-bottomed tidal channels separating the mangrove islands, analogous to the networks of channels in the Ten Thousand Islands. In some places, the ripples in the prehistoric channels and the tiny lenses of mud between the ridges were so perfectly preserved that they presented a complete geological record of each incoming and outgoing tide—events that happened in a matter of hours, 99 million years ago.

In addition, the geologists had found fine-grained beach deposits that they now realized lay between the remnants of individual mangrove islands, acting as barrier beaches along the seaward margin of the coast. Exactly the same kind of pattern can be seen in the Ten Thousand Islands of the Florida Gulf coast.

What Lacovara and Smith had not found, largely because they did not yet know what to look for, were substantial and persuasive organic remains of the mangrove-like shrubs and trees that had formed the islands in the first place. But that, too, would come.

"Just as Josh, on the first day, had the paleontologists on the team develop a search image of what dinosaur bones look like in Bahariya," Lacovara explains, "we needed a search image for what these plant remains would look like."

The first thing he began to recognize were vertical root structures, with the branching systems one might expect of any plant. He also saw densely packed vertical root casts characteristic of mangrove species. He had found such root systems in the rock above, below, and around the bones of the giant sauropod that the team had excavated south of el Dist, right in the quarry.

In addition, Lacovara noticed curious and widespread mats of root systems that spread horizontally, rather than vertically. Since the sediments of Gebel el Dist have been undisturbed by major geological forces, they remain in their original horizontal position. Therefore, these horizontal root systems had to have been in their original position as well. The vast majority of plants have vertical root systems, but thick, interwoven horizontal root masses are characteristic of mangroves.

The specific genera of plants that are called "mangrove" today did not exist 99 million years ago in the Cenomanian age of the Late Cretaceous. But a variety of mangrove-adapted plants have existed for at least 300 million years. What Lacovara wanted to discover was the specific kind of plant that had created the mangrove environment on the southern shore of the Tethys Sea in the big embayment he had begun calling the Bahariya Bight.

In time, Lacovara found it:[8] a treelike fern called *Weichselia reticulata,* which flourished from the middle Jurassic Period of the Mesozoic Era to the end of the Cenomanian age of the Late Cretaceous. Thus it was at its greatest extent of development at precisely the time that Stromer's lost dinosaurs and the 2000 expedition's giant sauropod roamed the shores of the Tethys Sea. A plant that was adapted to living in shallow saltwater environments, it may have grown to a height of twenty feet (based upon a type of mangrove fern, *Acrostichum,* that exists today). It had a trunk Lacovara describes as "ropy," potentially thick prop roots like today's mangroves, and a tasty, feathery, fernlike top—with exactly the kind of leafy fronds giant sauropods' teeth were designed to strip. A grove of *Weichselia* was, for all intents and purposes, a sauropod salad bar.

Lacovara found the distinctive petrified roots of *Weichselia* growing directly out of tidal sediments in the layer where the dinosaurs of Bahariya had been found—in fact, he found it in the very quarry in which the giant sauropod had been found in 2000.

"When I found that, I just stood there and stared, thinking, 'There it is at last,' " remembers Lacovara. "I almost cried."

Later, Lacovara would learn that Ernst Stromer had also found pieces of wood during his 1911 expedition, but Stromer had no context in which to place them and no real idea what they were.

"The woody part of *Weichselia* looks unrelated to its ferny top," says Lacovara, "and if you didn't know better, you'd think they were two different plants. Stromer didn't know better. Eventually, other researchers, using just the woody material that had been collected in Bahariya, named the plant *Paradoxopteris stromeri*—literally 'Stromer's puzzling-looking plant.' Much later it was found to be none other than *Weichselia.*

"*Weichselia,*" he explains, "has a lot of unusual, xeromorphic features—that is, features that make it possible to live in a water-short environment. It has water storage cells, which makes its leaves look like those of a succulent plant. It has cells that limit the rate of transpiration, so more of the water it holds stays inside. And the surface of the leaf has a thick, waxy cuticle, or skin, which further reduces water loss to the air."

All of which raises a question: Why does a plant living in what many scientists believe was the hot, humid climate of the Cenomanian age of the Late Cretaceous of North Africa, some 99 million years ago, need to be designed as if it were living in a desert? Answer: because it lives in salt water and acquires moisture by a very complex microfiltration system that filters salt out of the water in which it is immersed.

"The less water they lose," Lacovara says, "the less they have to manufacture, so they grow slowly and have to be very stingy with water. If you were to find a plant like this, with these xeromorphic features, you might think it was growing in an arid environment."

But not necessarily. Though they grow in water, the plants themselves might just as well be in a desert, because the fresh water that falls as precipitation in a mangrove swamp is largely unavailable to them. It falls into the salt water in which the plants stand, or quickly runs off the peaty soil that has accumulated in larger islands. Proof of this fact can be found today on many of the mangrove islands of Florida's Gulf coast, a region of abundant rainfall. Lacovara notes, "You can walk six or seven feet from the shore of a mangrove island in the Ten Thousand Islands and find cactus growing. Why? Because even though it rains something like sixty inches a year, the fresh water disappears. Fresh water exists, but in these paralic or intertidal zones, it might just as well not; it's not available."

It's something a lot of researchers have missed, says Lacovara: "I've seen academic papers where one author says *Weichselia* is a desert plant and another says it's a coastal plant, all because of its xeromorphic characteristics. And that may well be why some scientists think North Africa was an arid environment ninety-nine million years ago. But I don't think it was. At that time it was quite close to the equator. I think the evidence suggests it was very like the Gulf coast of the Florida

Everglades: hot, humid, lush—and yet functionally arid. That's not to say there is never any fresh water in mangroves. There certainly are some mangrove islands in the Ten Thousand Islands area, where freshwater springs and pools exist in their interior, and there you find a different and more varied mix of plants, even water lilies. But most of the plants you see are those that are well suited to an ephemeral, freshwater-short environment."

Over breakfast at El Beshmo Lodge one morning, Lacovara laid out his intertidal tropical-forest idea for the rest of the team. They were intrigued. Josh Smith thought for a moment and then said, "Dude, if this was a mangrove, you're a god!"

His enthusiasm was not due to Lacovara's having solved the problem of the missing beach sand, although having studied geology as an undergraduate himself, Smith was interested. The principal reason for his excitement was that he realized Lacovara had just provided another part of the answer to Stromer's riddle.

Stromer's riddle does not stop with the question "What were three species of giant carnivorous dinosaurs eating?" Part of that answer had been provided by the Bahariya Dinosaur Project's discovery of the giant sauropod, which, as Lacovara once put it, "could feed roughly five hundred thousand of your closest friends a nice, juicy quarter-pound burger." But there almost certainly had to have been other plant-eating dinosaurs besides that one and Stromer's *Aegyptosaurus,* just to keep the three species of carnivores alive. For that to be the case, the region where they all lived had to have been immensely productive ecologically, capable of producing immense amounts of foliage for large numbers of sauropods, and, though none have yet been discovered, other herbivores.

The most productive ecosystem on Earth today, producing by far the greatest amount of biomass per acre per year, is the tropical rain forest. However, rain forests are composed almost completely of angiosperms, flowering plants and trees, and in the Late Cretaceous, angiosperms were still just getting established. The second most productive ecosystem on Earth? Mangroves and salt marshes. Thus, in the Cenomanian age of the Late Cretaceous, when no rain forests are believed to have existed, mangroves and salt marshes probably were the most productive environments on the globe. An intertidal forest, a mangrove like the one on the Florida coast, at the interface between freshwater and saltwater worlds, would have been a nutrient-rich environment capable of supporting an astonishingly diverse array of flora and fauna. Mangroves are bursting with life.

"And what we've found in Bahariya bears this out," says Lacovara. "You've got a wide variety of plants, you've got huge carnivores, you've got ridiculously large herbivores, and then you've got all the other guys, the fish and the crocs and clams and snails and more."

Says Josh Smith, "The parallel with the Everglades mangroves is just stunning. Not only do the sediments look the same, but often the fauna does, too. The Everglades has turtles, Bahariya has turtles. The Everglades has crocodilians, Bahariya has crocodilians. The matchups are amazing. Except that the Bahariya versions of each of these species are much bigger. They're giants. Why is there so much diversity and why is everything so big? Because there's a lot of food there for everyone."

A mangrove environment has two primary ecological conditions. First, the seawater in which the plants live must be low-energy, or without waves and surf. "Plant a tree on the beach in, say, Oregon," says Ken Lacovara, "and it won't be there after one tide cycle."

Second, the temperature of the coldest month cannot average lower than 68 degrees Fahrenheit. A map of the Earth during the Cenomanian age of the Late Cretaceous, when temperatures were higher worldwide than they are today, would show a very broad global belt that fell within this temperature range, much broader than would be the case today. Superimpose upon that map the outline of the coasts at that time, many of which were epeiric and thus shallow and low-energy, and it becomes clear that the ecosystem requirements of mangrove-like intertidal forests—low-energy shorelines and warm temperatures—existed in many places around the globe.

"I suspect this was a huge biome," says Lacovara, "one that is only just beginning to be explored."

It seems certain that this biome was one in which giant sauropods, like the one discovered by the Baharyia Dinosaur Project, felt at home, though Ken Lacovara suspects they did not spend their entire lives there.

"They might have lived back on the mainland and only gone into the mangroves a couple of times a day to graze," he suggests. "Or they might have gone in a couple of times a year, during migrations. They might have used the shallow channels as an easy way to get up or down the coast. We have no way of knowing how much of their life cycle they spent there. But anything that an animal as big as a sauropod did would have been pretty deliberate, I suspect; you don't just take a walk in the mangrove for the hell of it if you weigh eighty tons."

Still, the world of a mangrove would have been a remarkably hospitable place for a large herbivorous dinosaur. The sand in the tidal channels is fine-grained and evenly sized, and Lacovara had calculations done that demonstrated that these channels would have made a firm walking surface for a heavy animal. Even at high tide (and tide ranges would have been slight in any event), the water is likely to barely have reached the animal's shins—plenty deep for a wide variety of even quite large fish, but a shallow wade for a giant sauropod.

As it lumbered along these watery trails, past dense thickets of *Weichselia*, forage would have been plentiful—"sauropod cafeterias left and right," as Jen Smith puts it.

Visibility also would have been excellent, since the animal's head, even unextended, would easily have cleared the tops of most of the forest foliage on the islands. Thus, both route taking and reconnaissance for predators would have been relatively easy. Whether a huge sauropod would volunteer to walk overland, through the thickets on the islands, is unclear. Footing might have been less certain than in the channels, but for a very large animal, small trees like *Weichselia* would hardly have been an obstacle to progress. But where there are herbivores, carnivores are certain to follow, as predators and as scavengers.

There is no way to say with any certainty how this gigantic sauropod met its end in Bahariya's mangroves. Given its sheer size, a single-predator attack or killing is probably extremely unlikely. Even a would-be attacker as large as any of the three carnivores Stromer discovered nearby would have been dwarfed by such a beast and might well have thought twice about attacking it.

Could several theropods have ganged up on the mild-mannered giant as it quietly munched *Weichselia* fronds? It has been proposed that smaller carnivores may have hunted in packs, but there is relatively little evidence of giant carnivores having done so.[9] An injured or diseased sauropod, however, may have been an easy mark for a lone hunter. A group of sauropods may have migrated through the mangrove, and one of them may have lagged behind and fallen prey. Or it could have expired alone from any of a number of causes and been found and fed upon by scavengers. Certainly the scattered nature of some of its bones would support such a theory, as would the discovery of a theropod tooth amid the sauropod's bones.

What does not seem either likely or possible is that the huge sauropod was what Josh Smith calls a "bloat and float." That is, an animal that died some great distance away, became bloated by decomposition, and floated to the location where it was eventually found, in the process distributing parts of its skeleton by wind or current to parts unknown. The environment in which the researchers believe it died makes that a virtual impossibility.

Ken Lacovara explains: "In some cases, a dinosaur will die and its remains can be transported long distances by wave action or a fast-moving stream or river. But there is absolutely no evidence in the sedimentological record where this sauropod was found of any of those conditions. The rock in which it was found is very fine-grained, the kind of grains that can only be deposited in a very calm, low-energy fluid environment. The channel would have been very shallow, perhaps a few feet deep at high tide. A body of water that is incapable of moving even fine-grained silt and mud is not going to move an eighty-ton dinosaur or even a two-hundred-pound dinosaur humerus. It's not going to happen. We may not know how that sauropod died, but we do know it died in place. It keeled over or was brought down, but it did not drift to the place where we found it."

And for precisely the same reasons—the gentle tidal flow of calm, silt- and sand-laden intertidal waters—the body of the enormous sauropod was, tide after tide, quickly surrounded and then covered by sediments.[10] The Tethys Sea receded, inundated the region, and receded again many times, judging by what the geologists can tell from the layers of the Gebel el Dist member of the Bahariya Formation, steadily building layer upon layer of sediment atop the now ancient corpse. Layers accumulated. Layers eroded. New layers were laid down and eroded away again. Tens of millions of years passed, and hundreds of feet of rock rose above it. Then, millions of years ago, the limestone layer that had been deposited by Eocene seas was exposed, bowed upward, and cracked. Wind and water widened the cracks and, over hundreds of thousands more years, scooped out a depression where water eventually emerged and human beings began to settle. Succeeding generations of settlers gave the oasis several names, the most recent of which was Bahariya. Explorers came and went, and one, a German, stayed long enough to find dinosaur bones, only to have them turned to dust by a man-made cataclysm more destructive than the pressure of tens of millions of years' worth of rock.

Then one day, in the winter of 1999, three scientists drove along the bottom of this depression, and one of them, a young paleontologist named Josh Smith, caught sight of something out of the corner of his eye, something that looked like dinosaur bone. They stopped. They marked the spot. And a year later they returned and began digging, eventually releasing the great beast from its ancient tomb.

MEMORIALS

They called it *Paralititan,* meaning "giant near the sea."

In so doing, the members of the Bahariya Dinosaur Project lifted the sauropod dinosaur they discovered beneath the oasis floor from the anonymity of deep time and gave it unique status: The tidal giant was a new genus and species of dinosaur, never before known.

There is a precise protocol to the naming of new dinosaurs, codified in the *International Code of Zoological Nomenclature* (ICZN) but actually dating back to the eighteenth century and Carolus Linnaeus's seminal work *Systema Naturae.* Every creature, Linnaeus proposed, shall have two names. The first name is the genus to which the creature belongs. The second is the name of its species, called a species epithet.[1] The name *Tyrannosaurus rex,* to choose perhaps the most well known, signifies that a species, *rex,* was found to belong to a previously unknown genus of theropods, *Tyrannosaurus.* When a paleontologist finds a dinosaur that subsequent research shows has already been identified and described, the specimen gets no new name. A rule called the principle of priority takes effect, and only the original name is recognized. Perhaps the second best-known dinosaur—a long-necked, long-tailed sauropod once called *Brontosaurus*—no longer exists by that name because it was found, years after it was discovered, to be identical to a previously named sauropod called *Apatosaurus*. The previous name had priority, and *Brontosaurus* disappeared as a distinct name. However, if a dinosaur is found that belongs to neither a known species nor a known genus, the scientists making the announcement get to create new names for both. And this was the case with *Paralititan*. But it did not happen overnight. It was three weeks after the completion of the 2000 field season that the fossils collected by the Bahariya Dinosaur Project team—those, at least, that Egyptian officials authorized to be shipped—arrived at the loading dock of the Academy of Natural Sciences in Philadelphia. It was many more months before fossil preparator Jason Poole and his crew had extracted most of the sauropod's remains from their plaster jackets and removed enough of the rock encasing the bones for the paleontology team to painstakingly identify the remains. Only after Matt, Peter, and Josh had measured, photographed, sketched, and documented the length, width, and depth, not only of each bone but of every bump and hollow in each bone, could they begin comparing the new fossils to similar bones of already identified sauropods.

By the late fall of 2000, they were reasonably certain they were dealing with a new genus and species. But by then, the 2001 field season was upon them, and they wanted to determine if there were more pieces of the huge creature's skeleton buried in the quarry rock before making a formal announcement. In that regard—though not in others—the 2001 field season was a disappointment. Jason Poole and Egyptian geologist Yousry Attia, along with a team of local laborers, spent weeks expanding the titanosaur site to seven times its original size but came up with nothing new of significance.

"The quarry was the size of an olympic swimming pool," recalls Josh Smith, "and it was empty." This finding may well reinforce the notion that *Paralititan* was scavenged after it died; team members suspect the other parts of the enormous sauropod's skeleton are not far away.

In other ways, the 2001 season (once again underwritten by the Bahariya Dinosaur Project's partner, Cosmos Studios, and filmed by MPH Productions) was successful. In addition to excavating far more of the immensely productive Jon's Birthday site, the team recovered a partial skeleton of a theropod dinosaur, gypsum-free and well preserved; the skeleton of another large animal as yet unidentified, though believed to be either another theropod or a large crocodilian; a large lower jaw, perhaps of the huge coelacanth *Mawsonia;* and many other fossils, including one thimble-size discovery that, if found to be from the Late Cretaceous, may prove "bigger" than *Paralititan.* Many of these specimens are still in Egypt.

While the team was in Egypt for the 2001 expedition, it completed a formal professional paper describing both the giant sauropod they had found in 2000 and—rare for such articles—the world in which the creature had lived.

After almost a year of research, the Bahariya Dinosaur Project paleontologists concluded that the sauropod was not only a new species but a new genus as well, positioned within the titanosaurid clade—just as Matt Lamanna had hypothesized the day the great beast's vertebrae were unearthed. It was time to give it a name.

As in Linnaeus's time, the language of species identification today is Latin—or occasionally "Latinized" English. This presents a bit of a problem, however, in the case of dinosaurs: They were unknown during the Roman Empire. Consequently, there are no ready-made Latin names to apply to them. So paleontologists have the honor, and the challenge, of creating new names. The ICZN permits scientists to create any name they choose when identifying a new species or genus, so long as it has not already been used. Grammatically, the species name, which comes second, modifies the genus name.

After a good deal of discussion, the members of the Bahariya Dinosaur Project chose the full name *Paralititan stromeri.* In a felicitous way it captured the driving passion of the entire group—to focus not just on the "trophy" of a new dinosaur discovery but equally on the discovery of the world in which that dinosaur had lived, its paleoenvironment. The new name presents, simultaneously, two hypotheses: first, that this is indeed a new species and genus deserving of a new name; and second, that this enormous beast (*titan*) was found in, and spent part of its life in, a tidal (*paralic*) environment. Thus, *Paralititan.*

But it is also a name that memorializes, that lifts two beings from the layered strata of paleontological and historical obscurity into the light of scientific and public recognition.

Once Jason Poole has completed the process of cleaning and preserving the bones in the fossil preparation laboratory of the Academy of Natural Sciences in Philadelphia, *Paralititan*'s remains will be physically memorialized when they are returned to Egypt. They will be displayed at the Cairo Geological Museum, along with many of the other fossils discovered by the joint American-Egyptian Bahariya Dinosaur Project in 2000 and 2001. This immense dinosaur—the second most massive dinosaur ever found anywhere in the world—will take its rightful place of honor in the land of its birth, life, and death. As team member Medhat Said has said of his own work on the project and of the importance of *Paralititan* to his country, "Egypt has given us so many blessings; now we can give it something back in thanks."

The other memorial can be found in *Paralititan*'s species epithet, *stromeri*. The species epithet of a newly discovered dinosaur can be almost any name its discoverer chooses. The team could have named it after the place in which it was found, Bahariya, or simply Egypt. They could have chosen a Latin word that signified the dinosaur's enormous size.

Instead, they chose *stromeri* to memorialize a remarkable and little known German paleontologist whose decades of exceptional and meticulous work on the fossils of Bahariya had been vaporized in the early hours of April 25, 1944, for no other reason than because the world had gone mad.

Despite his purported frailty, despite the loss of much of his family's fortune in the dark years following World War I, despite the threats of the Nazis, the loss of two of his sons in World War II, and the destruction of much of his life's work during that RAF bombing raid, Ernst Stromer endured. He was still writing and publishing scientific monographs well into his late sixties.

Appreciations written about him by fellow scientists cite the strength of his will as the explanation for his longevity. Perhaps. His surviving family members have another explanation, however. He was waiting for Wolfgang, the son simply referred to as "missing" at the Russian front during World War II, to return. Miraculously, on May 5, 1950, Wolfgang did. A physicist by education, he had been pressed by the Russians to help produce poison gas after his capture. When he repeatedly refused, he was held in various prison camps in Siberia for years. When he finally was released, his father, then eighty, was waiting for him. Stromer lived long enough to learn that Wolfgang and his new wife, Heidrun Rühle, would soon have a child, a daughter, Rotraut, the first girl to be born in Stromer's family in a century.

And then, on December 18, 1952, just a few days before Christmas, Ernst Freiherr Stromer von Reichenbach died.

He is buried in a family plot on the grounds of the castle, Grünsberg. Appropriately, for a man who spent so much time in the desert, the stone he is buried under is sandstone. Only his family name is chiseled upon the stone. There is no epitaph.

But there will be soon, in the place he seems to have loved almost as much as Grünsberg. In Cairo, in the halls of the Geological Museum, both he and the great titanosaur will at last become public history with the display of the bones of an immense dinosaur and a plaque in Latin that reads simply:

Paralititan stromeri.

PROLOGUE: DEATH AND RESURRECTION

1. Joshua B. Smith, Matthew C. Lamanna, Kenneth J. Lacovara, Peter Dodson, Jennifer R. Smith, Jason C. Poole, Robert Giegengack, and Yousry Attia, "A Giant Sauropod Dinosaur from an Upper Cretaceous Mangrove Deposit in Egypt," *Science,* Vol. 292, June 1, 2001, pp. 1704–6.

CHAPTER ONE: REAPING THE WHIRLWIND

1. Rick Gore, "Extinctions," *National Geographic,* Vol. 175, No. 6, June 1989, p. 669.
2. Tim Haines, *Walking with Dinosaurs* (London: Dorling Kindersley, 1999), p. 28.
3. Vincent Courtillot, "A Volcanic Eruption," *The Scientific American Book of Dinosaurs,* ed. Gregory S. Paul (New York: St. Martin's Press, 2000), p. 359.
4. "Deccan" means "southern" in Sanskrit; "trap" means "staircase" in Dutch.
5. Courtillot, "A Volcanic Eruption," p. 364.
6. Walter Alvarez and Frank Asaro, "An Extraterrestrial Impact," *The Scientific American Book of Dinosaurs,* p. 348.
7. Alvarez and Asaro, "An Extraterrestrial Impact," pp. 346 and 350.
8. Gregory Paul, "The Yucatan Impact and Related Matters," *The Scientific American Book of Dinosaurs,* p. 381.
9. It has even been suggested that the Deccan Traps volcanism was related to the Yucatan event, caused by shock waves from the impact that were transmitted through the Earth to the location where what is now India was then located on the globe.

10. Gore, "Extinctions," p. 692.

11. Richard Morris, *Cheshire: The Biography of Leonard Cheshire, VC, OM* (London: Viking, 2000), p. 23.

12. John Keegan, *The Second World War* (New York: Penguin Books, 1990), p. 420.

13. Ibid.

14. Ibid., p. 421.

15. Morris, *Cheshire*, p. 124.

16. Wilbur H. Morrison, *Fortress Without a Roof* (New York: St. Martin's Press, 1982), p. 224.

17. Ibid.

18. Earl Beck, *Under the Bombs: The German Home Front, 1942–1945* (Lexington, Ky.: University Press of Kentucky, 1986), p. 133.

19. Martin Middlebrook and Chris Everitt, *The Bomber Command War Diaries* (London: Midland Publishing, 1995), p. 499.

20. Until, that is, the Luftwaffe developed fighters with guns that could shoot upward, permitting German fighters to attack Lancasters from below.

21. Other diversionary raids were also under way: 27 Mosquito fighter-bombers attacked Düsseldorf, 165 planes made sweeps over the North Sea, and additional planes made mine-laying runs and attacked transport facilities in Chambly, France. In all, some 1,178 RAF sorties were flown that night, completely overwhelming German defense forces. Personal communication with Squadron Leader Rob Glover, RAF Lincoln and District Air Crew Association, August 2001.

22. Hans-Günther Richardi, *Bomber über München* (Munich: W. Ludwig Verlag, 1992), p. 240.

CHAPTER TWO: THE BONE-HUNTING ARISTOCRAT

1. The field journals of Ernst Freiherr Stromer von Reichenbach, written in his nearly impenetrable Sütterlin script, were recently donated by his family to the Paleontological Museum of the Bavarian State Collection in Munich. "Herr Leuchs" would appear to refer to Kurt Leuchs (1881–1949), a stratigrapher who taught at the University of Munich.

2. Many weeks later, toward the end of their now separate expeditions, Frau Leuchs apologized to Stromer for having caused difficulty, and she and Stromer appear to have parted friends.

3. Gene Gurney, *Kingdoms of the Middle East and Africa* (New York: Crown Publishers, 1986), p. 1.

4. Peter Mansfield, *A History of the Middle East* (New York: Penguin Books, 1991), p. 44.

5. Stromer made two expeditions to Egypt just after the turn of the century, one in the winter of 1901–2 and a second in the winter of 1903–4. He spent most of his time during these expeditions exploring for mammal fossils in the Fayoum Oasis, not far from Cairo.

6. Muhammad Ali's eldest son, Ibrahim, a remarkable military leader, died just months before Ali.

7. Mansfield, *A History of the Middle East*, p. 88.

8. Ibid., pp. 88–89.

9. Ibid., p. 100.

10. André Raymond, *Cairo* (Cambridge, Mass.: Harvard University Press, 2000), p. 324.

11. Cassandra Vivian, *The Western Desert of Egypt* (Cairo: American University in Cairo Press, 2000), p. 37.

12. Ibid., p. 38.

13. Ibid., p. 39.

14. Elwyn L. Simons and Tab Rasmussen, "Vertebrate Paleontology of Fayum: History of Research, Faunal Review and Future Prospects," *The Geology of Egypt*, ed. Rushdie Said (Rotterdam: A. A. Balkema, 1990), p. 629.

15. Ibid.

16. Vivian, *The Western Desert of Egypt*, p. 38.

17. W. D. Matthew, "Climate and Evolution," *Annals of the New York Academy of Sciences*, Vol. 24, February 18, 1915, pp. 171–318.

18. Simons and Rasmussen, "Vertebrate Paleontology of Fayum," pp. 628–29.

19. Harold L. Levin, *The Earth Through Time, Fifth Edition* (Orlando, Fla.: Harcourt Brace & Co., 1996), p. 6.

20. Simon Winchester, *The Map That Changed the World* (New York: Harper-Collins, 2001), p. 13.

21. Levin, *The Earth Through Time*, p. 11.

22. Personal interview, Hans-Dieter Sues, Royal Ontario Museum, Toronto, August 2001.

23. For the complete story of this historic achievement, a story as convoluted as the landscape it mapped, see Winchester, *The Map That Changed the World*.

24. Levin, *The Earth Through Time*, p. 14.

25. Michael Benton, "A Brief History of Dinosaur Paleontology," *The Scientific American Book of Dinosaurs*, ed. Gregory S. Paul (New York: St. Martin's Press, 2000), p. 11.

26. Winchester, *The Map That Changed the World*, p. 112.

27. Benton, "A Brief History of Dinosaur Paleontology," p. 17.

28. Except, of course, for mass extinctions.

29. Buckland—who, it will be remembered, named the first dinosaur, *Megalosaurus*—seems to have been compelled to taste his way through the ani-

mal and plant kingdom and the world of rocks. He could identify dirt and stone by its taste, and once famously debunked the recurrent "martyr's blood" at a European cathedral by dropping to his knees, licking the stain, and pronouncing it bat's urine.

30. From: Hans-Dieter Sues, "European Dinosaur Hunters," *The Complete Dinosaur*, eds. James O. Farlow and M. K. Brett-Surman (Bloomington, Ind.: Indiana University Press, 1997); and Benton, "A Brief History of Dinosaur Paleontology," pp. 10–44.

31. Richard Dehm, "Professor Dr. Ernst Freiherr Stromer von Reichenbach, Lebensdaten und Schriftverzeichnis," *Mitt. Bayer. Staatsamml. Paläont. Hist. Geol.*, No. 11, December 15, 1971, pp. 3–10.

32. *Lehrbuch der Paläozoologie. II. Teil: Wirbeltiere.* 325 S., B. G. Teubner, Leipzig, 1912.

CHAPTER THREE: UNEARTHING A LEGEND

1. It was another Western Desert explorer, the engineer, pilot, archaeologist, and reputed spy Count Ladislaus Edouard de Almas, who was the pivotal figure in Michael Ondaatje's extraordinary novel *The English Patient*.

2. Personal interview, Robert Giegengack, July 2001.

3. A "gebel" is a hill or ridge.

4. Thomas R. Holtz, Jr., "Classification and Evolution of the Dinosaur Groups," *The Scientific American Book of Dinosaurs*, ed. Gregory S. Paul (New York: St. Martin's Press, 2000), pp. 144–47.

5. Thomas R. Holtz, Jr., and Michael Brett-Surman, *Dinosaur Field Guide* (New York: Random House, 2001), pp. 76–77.

6. Three volunteers paid a total of $9,000 to participate as well.

CHAPTER FOUR: DRAGOMEN, FOSSILS, AND FLEAS

1. Stromer's personal journals, December 26, 1910, translated by Gisela Meckstroth.

2. John Ball and Hugh J. L. Beadnell, *Baharia Oasis: Its Topography and Geology* (Cairo: National Printing Department, 1903), p. 20.

3. Ibid., p. 37.

4. Cassandra Vivian, *The Western Desert of Egypt* (Cairo: American University in Cairo Press, 2000), p. 180.

5. G. A. Hoskins, *Visit to the Great Oases of the Libyan Desert* (London: Longman, Rees, Orme and Co., 1837), describing Kharga Oasis and quoted in Ball and Beadnell, *Baharia Oasis*, p. 17.

6. In fact, the dominant life-form in all three were single-celled organisms.

7. Elwyn L. Simons and Tab Rasmussen, "Vertebrate Paleontology of Fayum," *The Geology of Egypt*, ed. Rushdie Said (Rotterdam: A. A. Balkema, 1990), pp. 628–38.

8. And should have terrified Stromer as well, since the scorpions in this part of Egypt are among the most lethal in the world.

9. In his journal, Stromer incorrectly identifies this hill as Gebel Maghrafa, but at some later point he corrects his entries. That it is el Dist and not Maghrafa is made clear by his description of the hill and his reference to the page in the Ball and Beadnell survey of Bahariya that describes el Dist.

10. These measurements closely correspond to a femur and fibula sent to Munich in 1912. Stromer later identified them as belonging to a new carnivorous dinosaur he called *Bahariasaurus*.

11. Possibly the caudal, or tail, vertebra of a titanosaurid sauropod.

12. Certain titanosaurids are known to have had armor plates.

CHAPTER FIVE: THE ROAD TO BAHARIYA

1. John Ball and Hugh J. L. Beadnell, *Baharia Oasis: Its Topography and Geology*, (Cairo: National Printing Department, 1903), p. 18.

2. Farouk El-Baz, "Egypt's Desert of Promise," *National Geographic*, Vol. 161, No. 2, February 1982, pp. 190–220.

3. A. M. Allam, "A Regional and Paleoenvironmental Study on the Upper Cretaceous Deposits of the Bahariya Oasis, Libyan Desert, Egypt," *Journal of African Earth Sciences*, Vol. 5, No. 4, 1986, p. 407.

4. Ulf Thorweihe, "Nubian Aquifer System," *The Geology of Egypt*, ed. Rushdie Said (Rotterdam: A. A. Balkema, 1990), p. 606.

5. Ahmed Fakhry, *The Egyptian Deserts: Baharia Oasis, Vol. I* (Cairo: Government Press, 1942), p. 10.

6. A unit of agricultural land, equivalent to about one acre.

7. Ball and Beadnell, *Baharia Oasis*, p. 44.

CHAPTER SIX: FINDS AND LOSSES

1. Baron Franz Nopcsa, "Ergebnisse der Forschungsreisen Prof. E. Stromers in den Wüsten Ägyptens," *Abhandlungen der Bayerischen Akademie der Wissenschaften, Mathematisch-naturwissenschaftliche Abteilung* 30(4): 3–27, Pl. 1, 1925. Foreword by Ernst Stromer, translation by John D. Scanlon, Department of Zoology, University of Queensland, Brisbane, Australia, 2000.

2. Stromer's 1915 paper (see below) credits Markgraf for having excavated the specimens "out of a small hill, from a whitish gray to yellowish, clayey, gypsum-free sandstone, below a cover of 30 cm ferruginous sandstone and 1 m of hard clay."

3. J. B. Bailey, "Neural Spine Elongation in Dinosaurs: Sailbacks or Buffalobacks?" *Journal of Paleontology,* Vol. 71, No. 6, 1997, pp. 1124–46.

4. Ernst Stromer, "Ergebnisse der Forschungsreisen Prof. E. Stromers in den Wüsten Ägyptens. II. Wirbeltier-Reste der Baharîje-Stufe (unterstes Cenoman)." 3. Das Original des Theropoden *Spinosaurus aegyptiacus* nov. gen., nov. spec. *Abhandlungen der Königlichen Bayerischen Akademie der Wissenshaften, Mathematisch-physikalische Klasse* 28(3): 1–32, 1915.

5. D. A. Russell, "Isolated Dinosaur Bones from the Middle Cretaceous of the Tafilalt, Morocco," *Bulletin du Muséum national d'histoire naturelle, Paris,* Series 18, 2–3, 1996, pp. 349–402; and P. Taquet and D. A. Russell, "New Data on Spinosaurid Dinosaurs from the Early Cretaceous of the Sahara," *Comptes Rendus de l'Académie des Sciences, Paris: sciences de la terre et des planètes,* Vol. 327, 1998, pp. 347–53.

6. P. Sereno, A. L. Beck, D. B. Dutheil, B. Gado, H.C.E. Larsson, G. H. Lyon, J. D. Marcot, O.W.M. Rauhut, R. W. Sadlier, C. A. Sidor, D. D. Varricchio, G. P. Wilson, and J. A. Wilson, "A Long-snouted Predatory Dinosaur from Africa and the Evolution of Spinosaurids," *Science,* Vol. 282, No. 5392, 1998, pp. 1298–1302.

7. A. Charig and A. C. Milner, "Baryonyx, a Remarkable New Theropod Dinosaur," *Nature,* Vol. 324, No. 6095, 1986, pp. 359–61; and Charig and Milner, *"Baryonyx walkeri:* A Fish-eating Dinosaur from the Wealden of Surrey," *Bulletin of the Natural History Museum (Geology),* Vol. 53, No. 1, 1997, pp. 11–70.

8. D. M. Martill, A.R.I. Cruickshank, E. Frey, P. G. Small, and M. Clarke, "A New Crested Maniraptor Dinosaur from the Santana Formation (Lower Cretaceous) of Brazil," *Journal of the Geographical Society, London,* Vol. 153, 1996, pp. 5–8; A.W.A. Kellner and D. de A. Campos, "First Early Cretaceous Theropod Dinosaur and Paleobiological Implications, Based Upon a New Specimen from Texas," *Neues Jahrbuch für Geologie und Paläeontologie Abhandlungen,* Vol. 199, No. 2, 1996, pp. 151–66; and H.-D. Sues, E. Frey, and D. M. Martill, "The Skull of *Irritator challengeri* (Dinosauria: Theropoda: Spinosauridae)," *Journal of Vertebrate Paleontology,* Vol. 19, No. 3, 1999, p. 79A.

9. P. Sereno et al., "A Long-snouted Predatory Dinosaur from Africa and the Evolution of Spinosaurids."

10. H.-D. Sues et al., "The skull of *Irritator challengeri* (Dinosauria: Theropoda: Spinosauridae)."

11. *Spinosaurus* would finally take center stage many years later, as one of the "starring" villains in the movie *Jurassic Park III.*

12. Hagen Schulze, *Germany: A New History* (Cambridge, Mass.: Harvard University Press, 1998), p. 210.

13. Stromer foreword, "Ergebnisse der Forschungsreisen Prof. E. Stromers in den Wüsten Ägyptens," p. 2.

14. Ibid.

15. Although lizards, pterosaurs, birds, mammals, and lissamphibians (the term for all modern amphibians, including frogs, salamanders, and caecilians) are all known to have existed during the Late Cretaceous, Stromer found no evidence of these creatures in the Bahariya Formation.

16. Its species, *saharicus,* was originally referred to another genus, *Dryptosaurus,* named in 1927 from an isolated tooth found in Algeria.

17. In 1934, Stromer also referred an ilium (a pelvic bone) to *Carcharodontosaurus* but this report may be spurious. See O.W.M. Rauhut, "Zur systematischen Stellung der afrikanischen Theropoden *Carcharodontosaurus* Stromer 1931 und *Bahariasaurus* Stromer 1934," *Berliner Geowissenschaftliche Abhandlungen* Vol. 16, 1995, pp. 357–75.

18. P. C. Sereno, D. B. Dutheil, M. Iarochene, H.C.E. Larsson, G. H. Lyon, P. M. Magwene, C. A. Sidor, D. D. Varricchio, and J. A. Wilson, "Predatory Dinosaurs from the Sahara and Late Cretaceous Faunal Differentiation," *Science*, Vol. 272, 1996, pp. 986–91.

19. O.W.M. Rauhut, "Zur systematischen Stellung der afrikanischen Theropoden *Carcharodontosaurus* Stromer 1931 und *Bahariasaurus* Stromer 1934."

20. Sereno et al., "Predatory Dinosaurs from the Sahara and Late Cretaceous Faunal Differentiation."

21. Personal interview, Hans-Dieter Sues, Royal Ontario Museum, Toronto, August 2001.

22. Obituary written by Rudolf Richter, Stromer family papers.

23. Congratulatory note by Rudolf Richter, translated by Hans-Dieter Sues, Royal Ontario Museum.

24. From Frau Mathilde Weigert letter in Stromer's personal papers. Weigert also describes Stromer's protection of one particular Jewish friend, Frau Irma Heilbronner, until she finally was arrested.

25. Personal interview, Hans-Dieter Sues, August 2001.

26. Ibid.

CHAPTER SEVEN: SAND, WIND, AND TIME

1. An Australian labyrinthodont, *Koolasuchus,* is believed to be some 15 to 20 million years older.

2. Subsequently identified as belonging instead to the marine squamate genus *Simoliophis.*

CHAPTER EIGHT: THE HILL NEAR DEATH

1. Paul Morgan, "Egypt in the Framework of Global Tectonics," *The Geology of Egypt*, ed. Rushdie Said (Rotterdam: A. A. Balkema, 1990), p. 91.
2. Ibid., p. 98.
3. R. McNeill Alexander, "Biomechanics," *Encyclopedia of Dinosaurs*, eds. Philip J. Currie and Kevin Padian (San Diego, Calif.: Academic Press, 1997), p. 51.
4. John Ball and Hugh J. L. Beadnell, *Baharia Oasis: Its Topography and Geology* (Cairo: National Printing Department, 1903), pp. 50–51.
5. Ernst Stromer, "Die Topographie und Geologie der Strecke Gharaq-Baharije nebst Ausführungen über die geologische Geschichte Ägyptens," *Abh. Bayer. Akad. Wiss., math-phys., K1*, Vol. 26, No. 11., Abh. 78S, 1914.
6. Gamal Hantar, "North West Desert," *The Geology of Egypt*, p. 313.

CHAPTER NINE: SOLVING STROMER'S RIDDLE

1. The humerus is the upper bone of the arm, between the shoulder and elbow joints. The deltopectoral crest is a raised ridge on the craniolateral (forward and external) side of this bone, to which deltoid and pectoral muscles attach.
2. See Luis M. Chiappe and Lowell Dingus, *Walking on Eggs: The Astonishing Discovery of Thousands of Dinosaur Eggs in the Badlands of Patagonia* (New York: Scribner, 2001).
3. Barry Cox, R.J.G. Savage, Brian Gardiner, Colin Harrison, and Douglas Palmer, *Encyclopedia of Dinosaurs and Prehistoric Creatures* (New York: Simon & Schuster, 1999), p. 8.
4. Personal interview, Thomas R. Holtz, Earth, Life and Time Program, University of Maryland, September 2001.
5. Namely, phylogenetic systematics (also known as cladistics), evolutionary systematics, and phenetics (also called numerical taxonomy). Xiao-Chun Wu and Anthony P. Russell, "Systematics," *Encyclopedia of Dinosaurs*, eds. Philip J. Currie and Kevin Padian (San Diego, Calif.: Academic Press, 1997), p. 704.
6. Personal interview, Peter Dodson, University of Pennsylvania, September 2001.
7. Peter Dodson, "Distribution and Diversity," *Encyclopedia of Dinosaurs*, p. 187.
8. Personal interview, Thomas R. Holtz, September 2001.
9. See, for example, Louis Jacobs, "African Dinosaurs," *Encyclopedia of Dinosaurs*, pp. 2–4.

CHAPTER TEN: LOST WORLD OF THE LOST DINOSAURS

1. Peter Dodson, "Sauropod Paleoecology," *The Dinosauria,* eds. David B. Weishampel, Peter Dodson, and Halszka Osmólska (Berkeley, Calif.: University of California Press, 1992), pp. 402–7.

2. Ibid., p. 405.

3. Ibid., p. 404.

4. David Norman, "The Evolution of Mesozoic Flora and Fauna," *The Scientific American Book of Dinosaurs,* ed. Gregory S. Paul (New York: St. Martin's Press, 2000), p. 223.

5. Ibid.

6. P. Enos and R. D. Perkins, "Evolution of Florida Bay from Island Stratigraphy," *Geological Society of America Bulletin,* 90, pp. 59–83.

7. A. M. Allam, "A Regional and Paleoenvironmental Study on the Upper Cretaceous Deposits of the Bahariya Oasis, Libyan Desert, Egypt," *Journal of African Earth Sciences,* Vol. 5, No. 4, 1986, p. 407.

8. During the 2001 field season.

9. Intriguingly, one possible instance involves a relative of *Carcharodontosaurus.*

10. Lacovara found tidal deposits on Gebel el Dist more than forty inches thick that recorded each incoming and outgoing tide over only forty-three days.

EPILOGUE: MEMORIALS

1. George Olshevsky, "Naming the Dinosaurs," *The Scientific American Book of Dinosaurs,* ed. Gregory S. Paul (New York: St. Martin's Press, 2000), p. 131.

BIBLIOGRAPHY ════════════════════════════════════

PROLOGUE: DEATH AND RESURRECTION

Smith, Joshua B., Matthew C. Lamanna, Kenneth J. Lacovara, Peter Dodson,
    Jennifer R. Smith, Jason C. Poole, Robert Giegengack, and Yousry Attia, "A
    Giant Sauropod Dinosaur from an Upper Cretaceous Mangrove Deposit in
    Egypt," *Science,* Vol. 292, June 1, 2001.

ONE: REAPING THE WHIRLWIND

Alvarez, Walter, and Frank Asoro, "An Extraterrestrial Impact," *The Scientific
    American Book of Dinosaurs,* ed. Gregory S. Paul (New York: St. Martin's
    Press, 2000).
Beck, Earl, *Under the Bombs: The German Home Front, 1942–1945* (Lexington,
    Ky: University Press of Kentucky, 1986).
Courtillot, Vincent, "A Volcanic Eruption," *The Scientific American Book of Di-
    nosaurs,* ed. Gregory S. Paul (New York: St. Martin's Press, 2000).
Gore, Rick, "Extinctions," *National Geographic,* Vol. 175, No. 6, June 1989.
Haines, Tim, *Walking with Dinosaurs* (London: Dorling Kindersley, 1999).
Keegan, John, *The Second World War* (New York: Penguin Books, 1990).
Middlebrook, Martin, and Chris Everitt, *The Bomber Command War Diaries*
    (London: Midland Publishing, 1995).
Morris, Richard, *Cheshire: The Biography of Leonard Cheshire, VC, OM* (London:
    Viking, 2000).
Morrison, Wilbur H., *Fortress Without a Roof* (New York: St. Martin's Press,
    1982).

Paul, Gregory S., "The Yucatan Impact and Related Matters," *The Scientific American Book of Dinosaurs,* ed. Gregory S. Paul (New York: St. Martin's Press, 2000).

Richardi, Hans-Günther, *Bomber über München* (Munich: W. Ludwig Verlag, 1992).

TWO: THE BONE-HUNTING ARISTOCRAT

Benton, Michael, "A Brief History of Dinosaur Paleontology," *The Scientific American Book of Dinosaurs,* ed. Gregory S. Paul (New York: St. Martin's Press, 2000).

Dehm, Richard, "Professor Dr. Ernst Freiherr Stromer von Reichenbach, Lebensdaten und Schriftverzeichnis," *Mitt. Bayer. Staatsamml. Paläont. Hist. Geol.,* No. 11, December 15, 1971.

Gurney, Gene, *Kingdoms of the Middle East and Africa* (New York: Crown Publishers, 1986).

Lehrbuch der Paläozoologie. II. Teil. Wirbeltiere. 325 S., B. G. Teubner, Leipzig, 1912.

Levin, Harold L., *The Earth Through Time, Fifth Edition* (Orlando, Fla.: Harcourt Brace & Co., 1996).

Mansfield, Peter, *A History of the Middle East* (New York: Penguin Books, 1991).

Matthew, W. D., "Climate and Evolution," *Annals of the New York Academy of Sciences,* Vol. 24, February 18, 1915.

Raymond, André, *Cairo* (Cambridge, Mass.: Harvard University Press, 2000).

Simons, Elwyn L. and Tab Rasmussen, "Vertebrate Paleontology of Fayum: History of Research, Faunal Review and Future Prospects," *The Geology of Egypt,* ed. Rushdie Said (Rotterdam: A. A. Balkema, 1990).

Sues, Hans-Dieter, "European Dinosaur Hunters," *The Complete Dinosaur,* eds. James O. Farlow and M. K. Brett-Surman (Bloomington, Ind.: Indiana University Press, 1997).

Vivian, Cassandra, *The Western Desert of Egypt* (Cairo: American University in Cairo Press, 2000).

Winchester, Simon, *The Map That Changed the World* (New York: Harper-Collins, 2001).

THREE: UNEARTHING A LEGEND

Holtz, Thomas R., Jr., "Classification and Evolution of the Dinosaur Groups," *The Scientific American Book of Dinosaurs,* ed. Gregory S. Paul (New York: St. Martin's Press, 2000).

Holtz, Thomas R., Jr., and Michael Brett-Surman, *Dinosaur Field Guide* (New York: Random House, 2001).

FOUR: DRAGOMEN, FOSSILS, AND FLEAS

Ball, John, and Hugh J. L. Beadnell, *Baharia Oasis: Its Topography and Geology* (Cairo: National Printing Department, 1903).

Hoskins, G. A., *Visit to the Great Oases of the Libyan Desert* (London: Longman, Rees, Orme and Co., 1837).

FIVE: THE ROAD TO BAHARIYA

Allam, A. M., "A Regional and Paleoenvironmental Study on the Upper Cretaceous Deposits of the Bahariya Oasis, Libyan Desert, Egypt," *Journal of African Earth Sciences,* Vol. 5, No. 4, 1986.

El-Baz, Farouk, "Egypt's Desert of Promise," *National Geographic,* Vol. 161, No. 2, February 1982.

Fakhry, Ahmed, *The Egyptian Deserts: Baharia Oasis, Vol. I* (Cairo: Government Press, 1942).

Thorweihe, Ulf, "Nubian Aquifer System," *The Geology of Egypt,* ed. Rushdie Said, (Rotterdam: A. A. Balkema, 1990).

SIX: FINDS AND LOSSES

Bailey, J. B., "Neural Spine Elongation in Dinosaurs: Sailbacks or Buffalobacks?" *Journal of Paleontology,* Vol. 71, No. 6, 1997.

Charig, A., and A. C. Milner, "Baryonyx, a Remarkable New Theropod Dinosaur," *Nature,* Vol. 324, No. 6095, 1986.

Charig, A., and A. C. Milner, "*Baryonyx walkeri:* A Fish-eating Dinosaur from the Wealden of Surrey," *Bulletin of the Natural History Museum* (*Geology*), Vol. 53, No. 1, 1997.

Kellner, A. W. A., and D. de A. Campos, "First Early Cretaceous Theropod Dinosaur and Paleobiological Implications, Based Upon a New Specimen from Texas," *Neues Jahrbuch für Geologie und Paläeontologie Abhandlungen,* Vol. 199, No. 2, 1996.

Martill, D. M., A.R.I. Cruickshank, E. Frey, P. G. Small, and M. Clarke, "A New Crested Maniraptor Dinosaur from the Santana Formation (Lower Cretaceous) of Brazil," *Journal of the Geographical Society, London,* Vol. 153, 1996.

Nopsca, Baron Franz, "Ergebnisse der Forschungsreisen Prof. E. Stromer in den Wüsten Ägyptens," *Abhandlungen der Bayerischen Akademie der Wissenschaften, Mathematisch-naturwissenschaftliche Abteilung* 30(4): 3–27, Pl. 1, 1925. Foreword by Ernst Stromer, translation by John D. Scanlon, Department of Zoology, University of Queensland, Brisbane, Australia, 2000.

Russell, D. A., "Isolated Dinosaur Bones from the Middle Cretaceous of the Tafilalt, Morocco," *Bulletin de Muséum national d'histoire naturelle, Paris*, Series 18, 2–3, 1996.

Taquet, P., and D. A. Russell, "New Data on Spinosaurid Dinosaurs from the Early Cretaceous of the Sahara," *Comptes Rendus de l'Académie des Sciences, Paris: sciences de la terre et des planètes*, Vol. 327, 1998.

Schulze, Hagen, *Germany: A New History* (Cambridge, Mass.: Harvard University Press, 1998).

Sereno, P., A. L. Beck, D. B. Dutheil, B. Gado, H.C.E. Larsson, G. H. Lyon, J. D. Marcot, O.W.M. Rauhut, R. W. Sadlier, C. A. Sidor, D. D. Varricchio, G. P. Wilson, and J. A. Wilson, "A Long-snouted Predatory Dinosaur from Africa and the Evolution of Spinosaurids," *Science*, Vol. 282, No. 5392, 1998.

Sereno, P. C., D. B. Dutheil, M. Iarochene, H.C.E. Larsson, G. H. Lyon, P. M. Magwene, C. A. Sidor, D. J. Varricchio, and J. A. Wilson, "Predatory Dinosaurs from the Sahara and Late Cretaceous Faunal Differentiation," *Science*, Vol. 272, 1996.

Stromer, Ernst, "Ergebnisse der Forschungsreisen Prof. E. Stromers in den Wüsten Ägyptens. II. Wirbeltier-Reste der Baharîje-Stufe (unterstes Cenoman)." 3. Das Original des Theropoden *Spinosaurus aegyptiacus* nov. gen., nov. spec. *Abhandlungen der Königlichen Bayerischen Akademie der Wissenshaften, Mathematisch-physikalische Klasse* 28 (3): 1–32, 1915.

Sues, H.-D., E. Frey, and D. M. Martill, "The Skull of *Irritator challengeri* (Dinosauria: Theropoda: Spinosauridae)," *Journal of Vertebrate Paleontology*, Vol. 19, No. 3, 1999.

EIGHT: THE HILL NEAR DEATH

Alexander, R. McNeill, "Biomechanics," *Encyclopedia of Dinosaurs*, eds. Philip J. Currie and Kevin Padian (San Diego, Calif.: Academic Press, 1997).

Hantar, Gamal, "North West Desert," *The Geology of Egypt*, ed. Rushdie Said (Rotterdam: A. A. Balkema, 1990).

Morgan, Paul, "Egypt in the Framework of Global Tectonics," *The Geology of Egypt*, ed. Rushdie Said (Rotterdam: A. A. Balkema, 1990).

Stromer, Ernst, "Die Topographie und Geologie der Strecke Gharaq-Baharije nebst Ausführungen über die geologische Geschichte Ägyptens," *Abh. Bayer. Akad. Wiss., math-phys., K1*, Vol. 26, No. 11., Abh. 78S, 1914.

NINE: SOLVING STROMER'S RIDDLE

Chiappe, Luis M., and Lowell Dingus, *Walking on Eggs: The Astonishing Discovery of Thousands of Dinosaur Eggs in the Badlands of Patagonia* (New York: Scribner, 2001).

Cox, Barry, R.J.G. Savage, Brian Gardiner, Colin Harrison, and Douglas Palmer, *Encyclopedia of Dinosaurs and Prehistoric Creatures* (New York: Simon & Schuster, 1999).

Dodson, Peter, "Distribution and Diversity," *Encyclopedia of Dinosaurs and Prehistoric Creatures* (New York: Simon & Schuster, 1999).

Wu, Xiao-Chun, and Anthony P. Russell, "Systematics," *Encyclopedia of Dinosaurs*, eds. Philip J. Currie and Kevin Padian (San Diego, Calif.: Academic Press, 1997).

TEN: LOST WORLD OF THE LOST DINOSAURS

Dodson, Peter, "Sauropod Paleoecology," *The Dinosauria*, eds. David B. Weishampel, Peter Dodson, and Halszka Osmólska (Berkeley, Calif.: University of California Press, 1992).

Enos, P., and R. D. Perkins, "Evolution of Florida Bay from Island Stratigraphy," *Geological Society of America Bulletin*, 90.

Norman, David, "The Evolution of Mesozoic Flora and Fauna," *The Scientific American Book of Dinosaurs*, ed. Gregory S. Paul (New York: St. Martin's Press, 2000).

EPILOGUE: MEMORIALS

Olshevsky, George, "Naming the Dinosaurs," *The Scientific American Book of Dinosaurs*, ed. Gregory S. Paul (New York: St. Martin's Press, 2000).

ACKNOWLEDGMENTS ═══════════════════════════════

Every so often, a writer of nonfiction has the good fortune to come across an individual possessed of a great story waiting to be told. In writing this book, I had that good fortune five times over. I am deeply grateful to the five young scientists who established the Bahariya Dinosaur Project: Josh Smith, Matt Lamanna, Jennifer Smith, Ken Lacovara, and Jason Poole. Brilliant, devoted, determined, and absolutely delightful companions and collaborators, I offer them my deepest thanks for their time, their insights, and their abiding good humor and friendship. Thanks, too, go to University of Pennsylvania professors Dr. Robert Giegengack and Dr. Peter Dodson, whose wisdom and experience so enrich the project and this book.

I extend special thanks as well to Hans-Dieter Sues, vice president of collections and research at the Royal Ontario Museum in Toronto, whose knowledge of early European paleontologists in general and Ernst Stromer von Reichenbach in particular were especially valuable. In this regard I also am indebted to Stromer's surviving relatives, his granddaughter Rotraut Baumbauer and daughter-in-law Natalie Fryde.

Much of what is known about Stromer's activities in Egypt is contained in his own expedition journals. Written in a now-archaic form of German script called *Sütterlin,* the journals would have been useless had it not been for the knowledge, skill, patience, and keen eyesight of my translator, Gisela Meckstroth. Were it not for her, this would have been a very thin story indeed. Translation thanks are also due to Christine Stotz and Sven Sachs. For detailed historical information on the bombing of Munich, I am indebted to British Royal Air Force historian John Larder and to Rob Glover and William Williams of the RAF Lincoln and District Aircrew Association. For background information on Egyptian history I extend my appreciation to Cassandra Vivian and, for her always-rich private library, Joellyn Murphy.

For sharing their photo files and interview transcripts, and for bearing with me in the early days of researching this book, I thank Jim Milio and Melissa Jo Peltier of MPH Entertainment. For her support of the Bahariya Dinosaur Project and this book, I am deeply grateful to Ann Druyan of Cosmos Studios.

I owe the opportunity to write this book entirely to my agent and dear friend, Richard Abate of International Creative Management. I owe the joy I experienced in creating this book entirely to my editor at Random House, the talented and irrepressible Katie Hall. No writer was ever backed by a better or more enthusiastic team.

Finally, for her firm encouragement, abiding patience and steadfast friendship through an exceptionally difficult year, I express my deepest appreciation and affection to Kate Pflaumer.

Josh Smith, Matt Lamanna, Jennifer Smith, Ken Lacovara, and Jason Poole extend special thanks to the Egyptian members of the Bahariya Dinosaur Project, their field partners and friends Yousry Attia, Medhat Said Abdelghani, and Yassir Abdelrazik. They also thank Dr. Robert Giegengack and Dr. Peter Dodson, as well as Khyrate Soleiman and Ahmed Swedan of the Egyptian Geological Survey and Mining Authority for their help in making the Bahariya Dinosaur Project such a success. Thanks, too, are extended to field partners Jean Caton, Jean Lacovara, Kirk Johnson, Steve Kurth, Mandy Lyon, Doug Nichols, Allison Tumarkin, and Patti Kane-Vanni, as well as the staff of the El Beshmo Lodge in Bawiti. Thanks, too, are due to Dusti Lewars, Sherry Michael, Fred Mullison, Lisa Sachs, and all other preparators at the Philadelphia Academy of Natural Sciences who worked on *Paralititan,* as well as Ted Daeschler. And for their financial, moral, or professional support, they thank Jim Milio, Melissa Jo Peltier and Mark Hufnail of MPH Entertainment, Ann Druyan and Kent Gibson of Cosmos Studios, the late Emilie de Hellebranth, the Delaware Valley Paleontological Society, Barbara Grandstaff, Carrie Schweitzer, Bob Walters, Tess Kissenger, Bruce Mohn, Rainbow Studios, Hans-Dieter Sues, Helmut Mayr, Tom Holtz, Jerry Harris, You Hailu, Rud Sadleir, Rubén Martinez, Eric Buffetaut, Marc Weichmann, Oliver Rauhut, and the Interlibrary Loan staff at the University of Pennsylvania.

In addition, Josh Smith offers his personal thanks to his father and mother Raymond and Carol Smith, and to Luther, Tracey, Mike, Andrew, and, especially, Jen. Matt Lamanna thanks his father and mother, Carl Lamanna and Deborah Cooper, as well as Jon and other family and friends. Jennifer Smith thanks her mother and father, Noreen and Vincent Smith, along with Chris Williams and Josh. Ken Lacovara thanks his father, Bob Lacovara, and wife Jean, as well as Drexel University Dean Selcuk Guceri and University of Pennsylvania professors

Dr. Hermann Pfefferkorn and Dr. Edward L. Doheny. Jason Poole adds his thanks to his wife, Dusti, and daughter, Arielle. Not incidentally, Jason thanks his eighth grade biology teacher, Mrs. Rubin, for teaching him the scientific method and for opening his eyes to the world around him—an expression of appreciation for early teachers echoed by his other four American colleagues in the Bahariya Dinosaur Project.

INDEX

Abbas Ali, 26
Abdelghani, Medhat Said, 6, 84, 97, 123–25, 127–28, 134, 165, 204
Abdelrazik, Yassir
at Happy Fish site, 124, 127, 134
as member of Bahariya Dinosaur Project, 6, 84
as Muslim, 128
relationship with Matt Lamanna, 128
at South Sauropod site, 165
ablation lags, 101
Academy of Natural Sciences, Philadelphia, 41, 84, 87, 202, 204
Acrostichum, 193
Aegyptosaurus, 110, 130, 133, 160, 164, 166, 170, 171, 175, 195
Aegyptosuchus, 108
Africa. *See also* Egypt; Western Desert of Egypt
differences in dinosaur characteristics, 173–75
geology of, 109, 139–42, 145, 175, 183
Alexandria, Egypt, Stromer's 1910 arrival in, 21–25
Allosauroidea, 109
Allosaurus, 109
Alte Akademie, Munich
destruction of dinosaur collection, 7, 10, 19–20, 117

Stromer demands removal of paleontology collection during World War II, 116
American Museum of Natural History, New York, 30
amniotes, 57
ancient seas. *See* Tethys Sea
Andrews, C. W., 34
angiosperms, 184, 195
animal kingdom, 169
Apatosaurus, 202
Archaeopteryx, 40
Archean Eon, 73, 139
Argentina, 59
Argentinosaurus, 173, 175
Ashmolean Museum, Oxford, 38
asteroids, 9
Atlantic Ocean, 144, 145
Attia, Yousry, 6, 90, 124, 127, 128, 134, 165

Bahariasaurus, 110–11, 160, 175
Bahariasaurus ingens, 110
Bahariya Bight, 193
Bahariya Dinosaur Project. *See also names of sites and names of team members*
arrival in Egypt, 83
Cairo preparations for expedition, 84–86
description of findings, 4–5
four stages of expedition, 120

Bahariya Dinosaur Project (cont'd.)
  funding for expedition, 60, 62, 63,
    64
  lab work on collected fossils, 202,
    204
  letting off steam, 129, 132
  list of expedition members, 84–85
  member illnesses, 135–36
  partial summary of discoveries,
    157
  return to Josh's original site, 162–63,
    164
  sauropod and theropod discoveries,
    129, 130, 133, 136, 156–57, 164,
    166–68
  second season in the field, 202–3
  trip from Cairo to Bahariya Oasis,
    88–91
Bahariya Formation. See also Gebel el
    Dist
  division into distinct members, 147
  enduring questions about ancient
    ecosystem, 160–61, 195
  named by Stromer, 146–47
Bahariya Oasis. See also Bahariya Di-
    nosaur Project
  description of an oasis, 72, 94–95
  description of Bawiti, 98–99
  harshness of environment, 50
  Josh Smith's first visit to, 46–47, 51,
    54–57, 59, 60, 125
  as main focus of Stromer's 1910 ex-
    pedition, 32
  overview, 91–93
  population, 93–94
  pros and cons for new expedition,
    49–51
  second dinosaur extinction, 7, 10,
    19–20
  Stromer's arrival, 71–72
  Stromer's journey to, 69–71
  Stromer's work, 74–81
Ball, John, 29, 71, 75, 88, 93, 146
Barron, Thomas, 29
Baryonyx, 104
basement complex, 140, 144, 147
Bavaria, 106
Bavarian Academy of Sciences, 107,
    115

Bavarian State Collection of Paleon-
    tology and Historical Geology,
    Munich
  destruction of dinosaur collection, 7,
    10, 19–20, 117
  Stromer's appointment to, 105, 107,
    115
  before and during World War II,
    115, 116
Bawiti, Egypt, 55, 76, 91, 94, 98–99,
    176
beach sand, 154, 181, 185, 186–88,
    188, 190, 192
Beadnell, Hugh, 29, 30, 71, 88, 93,
    146
Bennett, Donald, 11, 13, 14, 15,
    19
Beurlen, Karl, 115–16
bird-hipped dinosaurs, 57
birds, relationship to dinosaurs, 58
bones. See also fossils
  condition of fossils, 85, 97
  found at South Sauropod site,
    165–66
  and gypsum, 79, 101, 133–34
  how they become fossils, 4
  jacketing for transport, 80, 126, 127,
    131, 135, 136
  premineralization, 159
  state of bones found by Stromer,
    101
  state of early bones found on
    Bahariya Dinosaur Project,
    100–101
  study of Happy Fish site, 124,
    127–28
  unlikelihood of finding fossils,
    157–59
  wrapping for transport, 80,
    126
brachoisaurids, 170, 171
Britain
  influence in Egypt, 25, 26, 27
  RAF bombing of Munich, Ger-
    many, 10–20
Broili, Ferdinand, 115, 116
Brontosaurus, 201–2
Buckland, William, 38, 40
Burgess Shale, 158

Cairo. *See also* Egypt
  Bahariya Dinosaur Project members
    in, 83–88
  Geological Museum, 84, 128, 204,
    205
  Stromer's 1910 arrival, 25, 27–31
Cambrian Period, 74, 158
camels and camel drivers, 32, 44, 68,
  72
Carcharodontosaurus, 109, 110, 111,
  160, 174–75, 177
Carcharodontosaurus saharicus, 109
Caribbean Sea, 9
carnivorous dinosaurs. *See also*
  theropods
  Bahariasaurus, 110–11, 160, 175
  Carcharodontosaurus, 109, 110, 111,
    160, 174–75, 177
  description, 58
  importance of Stromer's discoveries,
    160
  and mangrove hypothesis, 195, 198
  Spinosaurus, 103–5, 111, 160, 175
Caton, Jean, 84, 124, 165, 178
Cenomanian age, 147, 192, 193, 194,
  196
Cenozoic Era, 73, 75, 148
centipedes, 95
Cheshire, G. Leonard, 10–11, 14, 15,
  16, 17, 18–19
cladistics, 169–70
classes, defined, 168
Cleopatra (steamship), 21, 23–24
coastal environments, 143, 181–82,
  188, 191
coastal sedimentology, 83, 153, 187
Cochrane, Ralph, 11, 14, 15
coelacanths, 135, 203
comets, 9
continental drift, 142, 174
continents. *See also* Africa
  Africa and South America, 109,
    145, 175
  ancient, 139–41, 144, 145, 174
  consolidation, 8, 140–41
  rifts, 141, 144, 145, 175, 182
coordinates, 55, 56, 96, 162. *See also*
  global positioning system
  (GPS)

Cope, Edward Drinker, 41
Cosmos Studios, 5, 64, 202
creation, 35, 36
Cretaceous Period, 48, 75, 92, 127,
  142, 144, 145, 147, 148, 156,
  158, 161, 170, 173, 174, 175,
  181, 182, 183–85, 192, 193, 194,
  196
crocodilians, 136, 196, 203
crocodyliforms, 108–9
Cuvier, George Léopold Baron, 40

Dakhla Oasis
  as feature of Western Desert, 89, 91
  focus of Giegengack's research, 52,
    54, 60
Dambusters, Royal Air Force, 14
Darwin, Charles, 39, 169
de Hellebranth, Emilie, 64
de Lesseps, Ferdinand-Marie, 26
Deccan Traps, 8, 9
Delaware Valley Paleontology Society,
  64
*Delineation of the Strata of England
  and Wales with a Part of Scotland,*
  37
Deltadromeus, 110–11
deltopectoral crest, 171
Depression, 111
desert pavement, 89
Devonian Period, 74, 135
dinosaur bone layer concept, 125, 147
dinosaurs. *See also* Stromer's dinosaurs
  destruction of Alte Akademie col-
    lection, 7, 10, 19–20, 117
  developing search image, 97
  differences in species, 173–75
  family tree, 57–58, 170
  impact of landmass changes, 142,
    175
  importance of Stromer's discoveries,
    160
  manual-labor aspects of hunting, 80,
    85, 121–22, 126, 127, 131, 135,
    136
  mass extinctions, 8–10, 172
  naming protocol, 201–2
  origin of term, 57
  popular interest in, 39

dinosaurs *(cont'd.)*
    relationship of birds to, 58
    role in speculation about rifts be-
      tween continents, 142, 175
    speculation on their death and after-
      math, 3–4, 198–99
    Stromer's riddle, 160–61, 195
    teeth, 104, 124, 136
    types of, 57–58
    unlikelihood of finding, 157–59
    when they lived, 73, 74, 142, 170
diplodocids, 58
Dodson, Peter
    background, 61, 99
    description, 119–20
    examines Gebel el Fagga site, 164
    explains search image concept, 97
    at Happy Fish site, 122, 123, 124,
      127–28, 131, 134
    at Jon's Birthday site, 165, 176
    at Lizard Ridge site, 124, 136, 163,
      165
    as member of Bahariya Dinosaur
      Project, 6, 64, 84
    at press conference, 6
    at South Sauropod site, 167, 176
    and Stromer's riddle, 161
    and "supporting characters" concept,
      97, 176
    teaches field techniques to Egyptian
      colleagues, 128
Doldrums, 183
dorsal spines, 104, 105
Doyle, Arthur Conan, 180
dragoman, 67. *See also* el Hitu, Mo-
    hammad Hasranin
Drexel University, 83, 143
Druyan, Ann, 5, 64

Earth, age of, 35, 39, 72–73
earthquakes, 141, 145
Egypt. *See also* Western Desert of
    Egypt
    Cairo Geological Museum, 84, 128,
      204, 205
    geological endeavors, 50–51
    geology of, 139–42, 145
    German presence prior to World
      War I, 27–28

Giegengack's view of, 51–52
history of British and French influ-
    ence, 25–27
list of Egyptian Bahariya Dinosaur
    Project partners, 84–85
strengthening paleontological capac-
    ity, 128
Stromer's 1910 arrival in Cairo, 25,
    27–31
Upper Cretaceous Bahariya Forma-
    tion, 4–5
Egyptian Geological Survey (original),
    51, 70
Egyptian Geological Survey and Min-
    ing Authority (successor), 51, 84,
    128, 178
El Beshmo Lodge, 94, 135, 178,
    195
el Dist. *See* Gebel el Dist
el Hitu, Mohammad Hasranin, 67,
    68–70, 72, 77, 81
Entente Cordiale, 27
Eocene Epoch, 75, 148, 199
eons, defined, 73
epeiric seas, 183, 196
epicontinental seas. *See* epeiric seas
eras, defined, 73–74
erosion, 138, 144, 146, 148, 159, 181,
    182
Eudes-Deslongchamps, Jacques-
    Armand, 40
Euramerica, 140
evolution, 39, 169, 170
excavation
    discovery of one of Stromer's sites,
      125–26, 127, 133
    how it's done, 121, 122–23
    Josh's original site, 162
    Lizard Ridge site, 136
    as manual labor, 135
    South Sauropod site, 175,
      176–79
Exogyra, 154
extinctions. *See* mass extinctions

Farafra Oasis, 89, 91
Fawke, G. E., 16
Fayoum Oasis, 28, 29, 30, 34, 42, 68,
    69, 75, 91, 128

field work. *See also* excavation; jacketing fossils
  expedition stages, 120
  Peter Dodson as a teacher, 128
  second season of Bahariya Dinosaur
    Project, 202–3
  typical accommodations, 94
figured stones, 36–37
Flamenco Hotel, 84
flaser bedding, 187–88, 191
float, 59, 120, 121, 124
Florida Gulf coast. *See* Ten Thousand
  Islands, Florida
fluvial sedimentologists, 153
formations, 146–47
fossil water, 93
fossils. *See also* bones
  condition of bone, 85, 97
  and gypsum, 79, 101, 133–34
  handling in field, 127
  how they form from bones, 4
  jacketing for transport, 80, 126, 127,
    131, 135, 136
  second field season of Bahariya Di-
    nosaur Project, 202–3
  unlikelihood of finding, 157–59
Fraas, Eberhard, 30, 40–41
Fracastoro, Girolamo, 36
France, influence in Egypt, 25–26, 27
fraud paleontology, 113
funding, 60, 62, 63, 64

Gebel el Dist
  exploration by Bahariya Dinosaur
    Project, 96–97, 125, 143, 149–51,
    152, 154, 181, 185
  finding, 55–56, 59, 60, 96
  geology of, 138–39, 146, 147–48,
    186–87
  as member of Bahariya Formation,
    147
  problems with site, 161, 181
  Stromer's work there, 55, 78, 79–80
  survey coordinates, 55, 56, 96
Gebel el Fagga, 162, 163, 165, 173
Gebel Ghorabi, 74
Gebel Hafhuf, 77, 80
Gebel Hammad, 80
Gebel Maghrafa, 164–65

Gebel Mandisha, 77–78
genus, defined, 168, 201
geologic time scale, 72–74
Geological Museum, Cairo, 84, 128,
  204, 205
Geological Survey of Egypt, 28–29,
  29, 33, 103, 108
geology
  Bahariya Dinosaur Project tasks,
    143
  as detective work, 180–82
  dinosaur bone layer concept, 125,
    147
  Egyptian, 139–42, 145
  history of the field, 35–37
  landscape of Western Desert, 70–71
  study of Gebel el Dist, 138–39, 143,
    146, 147–51, 181, 185–87
  tools and techniques, 148–49
  as unraveling a story, 137, 139, 151,
    154–55
  vs. paleontology, 137
Germany
  Bavarian State Collection of Paleon-
    tology and Historical Geology, 7,
    10, 19–20, 105, 107, 115, 116,
    117
  buildup to World War I, 32–33,
    102
  destruction of Alte Akademie di-
    nosaur collection, 7, 10, 19–20,
    117
  influence in Egypt, 27–28
  Munich after World War I, 105–6
  RAF bombing of Munich on April
    24, 1944, 15–20
Gestapo, 112. *See also* Nazis
Ghauraq, Egypt, 69
Giegengack, Robert
  at Bahariya Depression, 55, 56, 59,
    60
  first trip to Western Desert, 46,
    51–54, 85, 125
  as member of Bahariya Dinosaur
    Project, 5–6, 84
  at press conference, 5–6
  view of Egypt, 51–52
Giganotosaurus, 58, 109, 175
Giza, 88

glaciers, 140
glauconite, 154, 186, 191
global positioning system (GPS),
    55, 96, 121, 149, 150, 151,
    164
Gondwana, 140, 144, 145
Göring, Hermann, 112, 116
gravel, 144, 152
Great Chain of Being, 169
Great Depression, 111
Great Exhibition of 1851, London,
    39
Great Sand Sea, 89
Great War, 32–33, 102, 105–6, 107,
    108, 112
Grünsberg castle, 106, 117, 205
Gubbio, Italy, 9
Gumaa (camel driver), 68, 72
gypsum, 79, 101, 133–34

Hadean Eon, 73, 139
Hadrosaurus, 41
Happy Fish site, 121–23, 124, 127–28,
    131, 134–35, 136
Harris, Arthur, 12, 13–14, 15
Hartmann, Herr, 67
Hennig, Willi, 169
herbivores, 58, 183–85, 195, 197. See
    also sauropods
Hess, Rudolf, 112
Himalayas, 145
Himmler, Heinrich, 112
Hitler, Adolph, 107, 112, 115, 116
Holmes, Sherlock, 180–81
Holtz, Thomas R., 160, 174
Hufnail, Mark, 63, 64
Hume, William Fraser, 29
Hutton, James, 35, 72

Iguanodon, 39
ilium (bone), 124, 130
India, 145
Institute of Egypt, 28
International Code of Zoological
    Nomenclature (ICZN), 201,
    203
intertidal sediments, 186, 188, 199
intertidal tropical-forest environment.
    See mangrove hypothesis

iridium, 9
Irritator, 104
ischium (bone), 124
Islam, 94, 98, 128
Ismail Ali, 26, 27, 28
isolated conical hill, 55, 78, 146. See
    also Gebel el Dist
Issawi, Bahay, 55

jacketing fossils, 80, 126, 127, 131,
    135, 136
Jon's Birthday site, 163, 165, 176,
    202
junk paleontology, 113
Jurassic Period, 74, 127, 145, 173, 174,
    193

Kane-Vanni, Patti, 84, 121–23, 127,
    134, 165
Kearns, R.S.D., 16
Kharga Oasis, 89, 91
kingdoms, defined, 169
Krupps von Bohlen und Halbach fam-
    ily, 63
Kurth, Steve, 84, 124, 125

labyrinthodonts, 127–28
Lacovara, Kenneth
    arrival in Cairo, 83–84, 86
    background, 83, 143
    as coastal sedimentologist, 83,
        153–54, 181
    discovers Happy Fish site, 121
    exploring Gebel el Dist, 96, 97, 118,
        119, 125, 129, 142, 146, 147, 148,
        149–50, 151, 152–53, 154, 181,
        185
    as geologist member of team, 83, 97,
        142, 153, 155, 164, 179
    mangrove hypothesis, 189–99
    as member of Bahariya Dinosaur
        Project, 5, 83–84, 129
    mystery of missing beach sand,
        180–82, 186, 187–88
    at press conference, 5
    at South Sauropod site, 176, 177,
        178
    vacation in Ten Thousand Islands,
        189

Lamanna, Matt
  background, 99–100
  discovers Jon's Birthday site, 163
  at Happy Fish site, 122, 135
  at ilium site, 124
  at Jon's Birthday site, 163, 165, 166
  at Lizard Ridge site, 123, 124
  as member of Bahariya Dinosaur
    Project, 5, 84, 90, 94–95, 129, 176
  at press conference, 5, 6
  relationship with Josh Smith, 47, 49,
    51, 59, 60, 99–100, 130
  relationship with Yassir Abdelrazik,
    128
  at South Sauropod site, 167, 168,
    170–71, 172
  at Stromer II site, 133
  and Stromer's riddle, 160
  view of Medhat Said Abdelghani,
    128
  view of Peter Dodson, 61
land bridges, 142
landmasses, ancient, 139–41, 144,
  174
Laurasia, 144, 145
Lawrence Livermore National Laboratory, 9
Leidy, Joseph, 41
Leonardo da Vinci, 36
Leuchs, Dr. and Frau, 24–25, 29, 31,
  64
Libycosuchus, 108, 136
Libypithecus markgrafi, 30, 43
Lightfoot, Bishop, 35
limestone
  cap atop Gebel el Dist, 138, 139,
    148
  in Western Desert, 91–92, 138, 139,
    148, 199
Linnaeus, Carolus, 168–69, 201, 203
lithosphere, 141
lithostratigraphic formations, 146–47
lizard fossils, 123
lizard-hipped dinosaurs, 57
Lizard Ridge site, 124, 136, 163, 165
lungfish, 97, 124, 163
Luxor, Egypt, 32, 44
Lyell, Charles, 37
Lyons, H. G., 29

Mameluke beys, 25, 26
mammal fossils, Stromer's interest in,
  33, 34, 75, 80
Mandisha, Egypt, 74–75, 76, 77, 120
mangals, defined, 191
mangrove, defined, 189
mangrove hypothesis, 189–99
mangrove swamp
  as productive ecosystems, 196
  root systems, 192, 193
  southern coast of Tethys Sea as,
    189
  Ten Thousand Islands as, 189, 191,
    192, 194
Mantell, Gideon, 40
maps, geological, 28, 29, 37, 48
Marginocephalia, 58
marine creatures, 144
marine sedimentologists, 153
marine sediments, 186, 188, 189, 191
Markgraf, Richard
  accompanies Stromer to Egyptian
    sites, 43, 44
  background, 30–31
  death of, 103
  discovery of one of his sites in 2000,
    125–26, 127, 133
  further explorations on behalf of
    Stromer, 81, 160
  Libypithecus markgrafi named in
    honor of, 30, 43
  Markgrafia libica named in honor of,
    30
  Moeripithecus markgraf named in
    honor of, 30
  problems shipping dinosaur bones to
    Stromer, 102–3
  unable to accompany Stromer to
    Bahariya Oasis, 44–45
Markgrafia libica, 30
Marsh, Othniel Charles, 41
Maslim, Mohammad, 68, 69, 72, 77,
  78
mass extinctions, 7–9, 74, 141, 172
Matthew, William Diller, 33, 108, 173,
  174
Mawsonia libyca, 135, 203
Mediterranean Sea, 145
Megalosaurus, 38

Mesozoic Era
 changes during, 142, 144–45, 146
 Cretaceous Period, 48, 75, 92, 127,
  142, 144, 145, 147, 148, 156,
  158, 161, 170, 173, 174, 175,
  181, 182, 183–85, 192, 193, 194,
  196
 defined, 73, 74, 142
 Jurassic Period, 74, 127, 145, 173,
  174, 193
 plant life, 183–84
 Triassic Period, 127, 144, 145, 170,
  173, 174
Microraptor, 58
Milio, Jim, 64, 130
Moeripithecus markgrafi, 30
Mongolia, 158
Mount Vesuvius, 158
MPH Entertainment, 62, 63, 64, 130,
 202
mud pits, 158
mudstone, 152–53, 176, 191
Muhammad Ali, 26
Muhammad (camel driver), 44
multicelled lifeforms, 73
Munich, Germany
 after World War I, 105–6
 Bavarian State Collection of Paleontology and Historical Geology, 7,
  10, 19–20, 105, 107, 115, 116,
  117
 destruction of Alte Akademie dinosaur collection, 7, 10, 19–20,
  117
 RAF bombing on April 24, 1944,
  15–20

naming of dinosaurs, 201–2
naming of *Paralititan stromeri,* 201,
 203–5
Napoléon Bonaparte, 25
Nasser, Gamel Abdel, 92
National Socialist German Workers'
 Party, 15, 107, 112
Natrun Valley. *See* Wadi el Natrun
Natural History Museum, London,
 108
natural selection, 39
Nazis, 15, 112, 114

Nile cataracts, 139
Nubian aquifer, 92–93
Nuremberg, as home of Stromer,
 22–23, 106

oases, 72, 91–93. *See also* Bahariya
 Oasis; Fayoum Oasis
Occam's Razor, 190
ocean floors. *See* seafloor
Oraan (camel driver), 32
orders, defined, 168
origin of life, 73
Origin of Species, The, 39, 169
ornithischians, 40, 57–58
ornithopods, 58, 184, 185
Osborn, Henry Fairfield, 30
Ottoman Empire, 25, 33
Owen, Richard, 40, 57

paleontology
 in American West, 41–42
 early field paleontologists, 40–41
 field accommodations, 94
 funding of expeditions, 60, 62,
  63–64
 history of the field, 34, 37–42
 strengthening Egyptian capabilities,
  128
 tools and techniques, 85, 121–22
 vs. geology, 137
paleosols, 191
Paleozoic Era, 73, 74, 135, 140
Pangaea, 8, 140–41, 142, 144, 174,
 182
Panthalassa, 141
Paradoxopteris stromeri, 193
paralic sediments, 188, 203
Paralititan stromeri
 announced in *Science* magazine, 5
 bone unveiled at press conference, 6
 naming of, 201, 203–5
Patagonia, 109, 158
Pathfinders, Royal Air Force, 13, 14,
 18
Peltier, Melissa Jo, 64
Penn (University of Pennsylvania), 5,
 41. *See also* Dodson, Peter;
 Giegengack, Robert; Lamanna,
 Matt; Smith, Joshua

periods, defined, 74
Permian Period, 74, 140
Peyer, Bernhard, 108
Phanerozoic Eon, 73–74, 158
plant kingdom, 169
plaster of paris, 85, 134. *See also* jacketing fossils
plate tectonics, 141–42, 144
plesiosaur bones, Stromer findings, 78, 81, 109
Plot, Robert, 38
polarity shifts, 141
Pompeii, 158
Poole, Jason
    background, 86–88
    as bone preparator, 126, 202, 204
    critter warnings, 95–96
    at ilium site, 124
    lab work on fossils collected by Bahariya Dinosaur Project, 202, 204
    as member of Bahariya Dinosaur Project, 5, 84, 129
    original humerus discovery, 156–57
    at press conference, 5, 6
    return to Josh's original site, 162–63, 164
    South Sauropod humerus discovery, 167–68, 170, 171, 172
    at South Sauropod site, 165–68, 176, 177
Precambrian, 73
predators. *See* carnivorous dinosaurs
principle of cross-cutting relationships, 37
principle of inclusion, 37
principle of original horizontality, 34
principle of original lateral continuity, 34
principle of superposition, 34
Principles of Geology, 37
Proterozoic Eon, 73, 140
pubis (bone), 124

quicksand, 158

RAF bombing of Munich, Germany, 10–20
rain forests, 195

Rayyan Valley, Egypt, 69
Red Sea Hills, 139
Reichstag, 112
Rennebaum, Elizabeth, 43, 44, 82, 107, 117
Rennebaum, Johann Adam, 67
Rennebaum family, 43, 82
reptiles, 57
Richter, Rudolf, 113
ridges, underwater, 182
rifts, 141, 144, 145, 175, 182
Rijksmuseum, Leiden, 42
rock, how it is transported in nature, 152–53
Rohlfs, Gerhard, 28, 32
Royal Air Force bombing of Munich, Germany, 10–20
Royal Geographical Society of Egypt, 28
Royal Ontario Museum. *See* Sues, Hans-Dieter
Rühle, Heidrun, 205
Russell, Dale, 111

Sadat, Anwar, 92
Sahara Desert, 88. *See also* Western Desert of Egypt
Said, Medhat. *See* Abdelghani, Medhat Said
Said Ali, 26
sand, 144, 146, 152, 180, 181, 188. *See also* beach sand
sandstone, 79, 153, 154, 191
sandstorms, 118–19, 126, 135, 137, 158
Saurierlagen, 125, 147
saurischians, 40, 57, 58
sauropods, see also *Paralititan stromeri*
    Aegyptosaurus, 110, 130, 133, 160, 164, 166, 170, 171, 175, 195
    Apatosaurus, 202
    in Bahariya, 58, 175
    Bahariya Dinosaur Project discoveries, 130, 133, 136, 156–57, 164, 166–68
    Brontosaurus, 201–2
    and cladistics, 170
    and mangrove hypothesis, 193, 195, 197–98

sauropods (cont'd.)
    naming of Bahariya Dinosaur Project find, 203–5
    role in speculation about rift between Africa and South America, 175
    speculation on their death and aftermath, 3–4, 198–99
    speculation on what they ate, 183–85, 193, 195, 197–98
    Stromer finds, 110, 160
Schweinfurth, Georg, 28, 29, 30, 33, 34, 43, 66
Science, 4–5
scorpions, 95
sea levels, 144, 146, 182, 187, 188
seafloor, 141, 144, 182, 188
search images, 97, 192
sedimentary geology, 151–55
Seeley, Harry Govier, 40
Senckenberg Museum, Frankfurt am Main, 113
Sereno, Paul, 110, 111
shallow seas, 182–83, 196
Shannon, David J., 16
Sheppards Hotel, 67
Sigilmassasaurus, 111
silt, 144, 146, 152, 188
Silurian Period, 74, 140
Simoliophis, 109, 123, 136
Simons, Elwyn, 128
single-celled organisms, 73
Sinraptor, 109
sites. See Gebel el Dist; Happy Fish site; Jon's Birthday site; Lizard Ridge site; South Sauropod site; Stromer I site; Stromer II site
Siwa Oasis, 89, 91
Smith, Jennifer
    in Cairo, 85, 86
    description of Bawiti residents, 98–99
    discovers Happy Fish site, 121
    exploring Gebel el Dist, 96–97, 119, 125, 129, 146, 147, 149, 150–51, 152, 153, 154, 181
    on first Bahariya expedition, 55, 56, 59, 60, 125
    as fluvial geologist, 153, 181
    as geologist member of team, 97, 142, 154–55, 164, 179
    on geology vs. paleontology, 137
    as member of Bahariya Dinosaur Project, 5, 84, 129
    at press conference, 5
    response to Lacovara's mangrove hypothesis, 190
    and Robert Giegengack, 52–53
    at South Sauropod site, 176, 178
Smith, Joshua
    background, 46–49
    and Dodson, 61
    as expedition leader, 85, 95, 96, 97–98, 120, 121, 125, 129, 136, 142, 157, 161–62, 175
    first trip to Western Desert, 46–47, 51, 54–57, 59, 60, 125
    at ilium site, 124
    as member of Bahariya Dinosaur Project, 5, 84, 90, 129
    origin of plan to find Stromer's lost dinosaurs, 47–49
    at press conference, 5–6
    relationship with Matt Lamanna, 47, 49, 51, 59, 60, 99–100, 130
    response to Lacovara's mangrove hypothesis, 195
    return to his original site, 162–63, 164
    and Robert Giegengack, 51, 54
    South Sauropod humerus discovery, 167–68, 170, 171, 172
    at South Sauropod site, 165–68, 176, 177
    at Stromer II site, 133
    talk at Society of Vertebrate Paleontology meeting, 62–63
Smith, William, 37, 147
snakes, 95
Society of Vertebrate Paleontology, 62
Soleiman, Khyrate, 84
South America, 59, 109, 158, 173, 175
South Sauropod site, 165–68, 175, 176–79
Soviet Republic of Bavaria, 106

species, defined, 168, 201
species epithet, 201, 204–5
spiders, 95
spinosaurids, 103–5
Spinosaurus, 103–5, 111, 124, 136, 160, 175
Spinosaurus aegyptiacus, 103–5
Spinosaurus B, 111
Spinosaurus maroccanus, 104
spoils piles, 125
stegosaurs, 58
Steindorff, Georg, 29, 33
Steno, Nicholas, 34–35
Stomatosuchus, 108–9
storms. *See* sandstorms
stratigraphy, 146
Stromer, Ernst. *See also* Stromer's dinosaurs
    announces discovery of *Spinosaurus aegyptiacus*, 103–5
    arrival in Alexandria, 21–25
    arrival in Bahariya, 76
    arrival in Cairo, 25, 27–31
    budget for his 1910 expedition, 63–64
    death of, 205
    description, 21
    destruction of his Alte Akademie dinosaur collection, 7, 10, 19–20, 117
    dinosaur identifications, 108–11
    discovery of one of his sites in 2000, 125–26, 127, 133
    early influences on, 28
    extent of his findings, 160
    financial problems after World War I, 107, 108, 126
    finds first large dinosaur bones at Gebel el Dist, 79–80
    full name, 22
    importance of his discoveries, 174–75
    journals of, 42
    journey home from Egypt in 1911, 82
    longevity of, 204–5
    names Bahariya Formation, 146–47
    preparations for 1910 Bahariya expedition, 66–68

    problems receiving dinosaur bones from Egypt, 102–3, 106, 107–8
    resistance to Nazis, 114–17
    Richter's appreciation of, 113–14
    riddle in his findings, 160–61, 195
    as student, 41
    wife and children, 117, 205
    writes monographs, 103
Stromer, Gerhard, 117
Stromer, Rotraut, 205
Stromer, Ulman, 117
Stromer, Wolfgang, 117, 205
Stromer I site, 126, 129, 131
Stromer II site, 133, 134, 136
Stromer's dinosaurs
    Aegyptosaurus, 110, 130, 133, 160, 164, 166, 170, 171, 175, 195
    Bahariasaurus, 110–11, 160, 175
    Carcharodontosaurus, 109, 110, 111, 160, 174–75, 177
    Spinosaurus, 103–5, 111, 160, 175
Suchominus, 104
Sues, Hans-Dieter, 47–48, 50, 62, 99, 104, 114, 116, 160, 185
Suez Canal, 26, 27
supercontinents, 141, 142, 144
survey coordinates, 55, 56, 96, 162. *See also* global positioning system (GPS)
System Naturae. See Linnaeus, Carolus
systematics, 168

taphonomic data, 123
taxa, defined, 168
teeth, dinosaur, 104, 124, 136
Ten Thousand Islands, Florida, 189, 191, 192, 194
Tethys Sea, 145–46, 148, 154, 183, 186, 187, 189, 199
tetrapods, 57
Tewfik Ali, 27
Theory of the Earth, 35, 72
theropods. *See also* carnivorous dinosaurs
    Bahariasaurus, 110–11, 160, 175
    in Bahariya, 60, 111, 124, 160
    Bahariya Dinosaur Project discoveries, 129, 130, 136

theropods (*cont'd.*)
  Carcharodontosaurus, 109, 110, 111, 160, 174–75, 177
  Carcharodontosaurus saharicus, 110, 111
  description, 58
  excavated from Bahariya Dinosaur Project, 203
  importance of Stromer's discoveries, 160
  and mangrove hypothesis, 198
  Spinosaurus, 103–5, 111, 160, 175
Third Reich. *See* Nazis
Thyreophora, 58
tidal flats, 186, 188
titanosaurs, 3–4, 58, 110, 172–73, 203
transgressive sequences, 187
Treaty of Versailles, 106
Triassic Period, 127, 144, 145, 170, 173, 174
Triceratops, 58
tropical rain forests, 195
Tumarkin, Allison
  at Happy Fish site, 122
  at ilium site, 124
  journal observations, 131
  as member of Bahariya Dinosaur Project, 84
  at South Sauropod site, 165, 167, 178
turtles, 109, 136, 196
type sections, 146
type specimens
  Aegyptosaurus baharijensis, 110
  Bahariasaurus ingens, 110
  search for *Spinosaurus,* 124
  Tyrannosaurus rex, 103, 109, 110, 160, 175, 201

unconformities, 153, 181, 188
underwater ridges, 182
University of California at Berkeley, 9
University of Maryland, 160
University of Munich, 42, 107
University of Pennsylvania, 5, 41. *See also* Dodson, Peter; Giegengack, Robert; Lamanna, Matt; Smith, Joshua

University of Zurich, 108
Upper Cretaceous Bahariya Formation, 4–5
Ussher, James, 35
Uweinat Mountains, 139

Vertebrata, 57, 73
vertebrate paleontology. *See* paleontology
Vinac, 127, 131
volcanic activity, 8, 141, 144, 145, 182
von Heune, Friedrich, 40
von Hindenburg, Paul, 111–12
von Meyer, Hermann, 40
von Zittel, Karl Alfred, 28, 41, 42

Wadi el Natrun, 29, 31, 33, 42–44
Walther's Law, 186
water, in oases, 91, 92–93
water boxes, 32
wavy bedding, 188
Weichselia reticulata, 193–94, 197
Wentworth Scale, 152
Western Desert of Egypt. *See also* Bahariya Formation; Bahariya Oasis
  description, 88–91
  explorations by others prior to Stromer, 28, 29
  Giegengack's field work, 51–57
  Josh Smith's first visit to, 46–47, 51, 54–57, 59, 60, 125
  Rohlfs' expedition, 28
  sandstorms, 118–19, 126, 135, 137, 158
  six oases, 91–93
  as Stromer's research focus, 23
Wilhelm, Kaiser, 32–33
William of Occam, 190
windstorms. *See* sandstorms
Winters, Scott, 61–62
Woodward, Arthur Smith, 108
World War I. *See* Great War
World War II
  destruction of Alte Akademie dinosaur collection, 7, 10, 19–20, 117
  fate of Bavarian State Collection of Paleontology and Historical

Geology, 7, 10, 19–20, 105, 107, 115, 116, 117
Hitler and the Nazis, 15, 107, 112, 114, 115, 116
RAF bombing of Munich, Germany, 10–20

wrapping fossils for transport, 80, 126, 127, 131, 135, 136

Young, George, 38
Yucatán, 9

ABOUT THE TYPE

This book was set in Caslon, a typeface first designed in 1722 by William Caslon. Its widespread use by most English printers in the early eighteenth century soon supplanted the Dutch typefaces that had formerly prevailed. The roman is considered a "workhorse" typeface due to its pleasant, open appearance, while the italic is exceedingly decorative.